A Univ of Magical Things

A Cosmic View Beyond the Myths of Religion and Scientific Materialism

David A. Yeats

outskirts
press

To Sheri

My Companion and Best Friend, My Love,
Who the universe has so sweetly offered to me as a partner in this lifetime.
How can I ever express the boundless gratitude I feel?

To My Children—David, Lara, Michael, Jeremy, and Julian—
who I love with all my heart.

Table of Contents

PART TWO: THE POINT

"The Universe Is Full of Magical Things
Patiently Waiting for Our Senses to Grow Sharper"

—William Butler Yeats[1]

1   This quotation has been attributed to (my ancestor) William Butler Yeats, and also to Bertrand Russell and the playwright Eden Phillpotts. It was probably Phillpotts, but for our purposes... close enough!

INTRODUCTION

I remember as a kid, I was always wondering about this world we live in. I often used to lie on the grass on a humid, sunny, summer day and watch the green blades in their silent, imperceptible growth, squeeze the snapdragons in imagined conversation, smell the air as the summer breeze blew through the maple leaves, and follow the erratic antics of robins and orioles, caterpillars, lady bugs, wasps, and ants— and I'd ponder. The vivid, living world around me seemed such a mystery: active, alive, and miraculous!

At night, I'd be entranced by lightening bugs and the background steady state sounds of the crickets. Did they ever sleep? I'd look up at that same sky that other humans have done for eons, and I'd feel the same mystery and awe they did, and I'd wonder, just as every pair of eyes has ever done: Where am I? Who are we? What am I looking at? What is *real?*

When I ask that question, I find myself remembering a story about my grandson, Andrew, who was five at the time when he came for a visit to Colorado. He lived with his family in L.A., and hadn't been out in nature very much. Andrew's grandma and I took him and his sister Natalia on a day trip to Rocky Mountain National Park, to the Fern Lake trail, for a hike. The Fern Lake trail parallels the Big Thompson Creek, near the headwaters of what would become the Big Thompson River. It's a beautiful trail with big views of the snow-capped mountains. We saw the water rushing by over glistening rocks, and lush greenery and mountain meadow flowers blooming on an exquisitely beautiful early summer day. We passed a pond with a few ducks playing on it. We were beholding a world well beyond the world of Disney, so familiar to our screen-savvy grandkids.

Suddenly, Andrew turned around, wide-eyed, and asked in amazement, "Grandma, is this *real!?*"

Is this *real?*

The natural world is so awe-inspiring, it begs the question! Andrew, in that moment, was as stunned by the force that is nature as I had been when, mesmerized, as a child, I studied those busy insects.

Is this *real?*

What follows in this book will, likely, challenge important beliefs our culture has taken for granted to be true. As a citizen of our western culture, you may be quite tempted to ask the same question when you read what is written here. *Is this real?* I will articulate a significant challenge to our current cultural scientific beliefs about what is real. However, there is no intention to cause anyone angst, or to be polemic. It is only another lens, another view to consider. Our culture naturally prefers we live our lives and think of our world in a way that maintains the view we've engineered in our traditions and civilization, but what if that entire view is based on some false assumptions?

Samuel Veissiére observed that "Prior learning prompts us to see the world as we expect it to be not as it is…Once a cultural schema is in place, and the cues around us look right enough, we are blind to details and context: we automatically process information as we expect it to appear, without investing any mental effort" (Veissiére, Aug 2017, p. 66). Our experiences tend to lead us to conclusions that are consistent with those experiences. But that often produces false assumptions and false conclusions.

Herein lies the core challenge, I believe, in allowing an open stance toward what is written in these pages. Our cultural schema has reinforced that we look at the world in a certain way—about all subjects, but particularly, for our purposes here, in the realms of science and spirituality. However, what if that patterned and culturally orthodox representation of what the world is is based in flawed assumptions?

I believe it is. So do many others. In fact, there are many conscientious, thoughtful, think-outside-the-box folks in our world who, even though they are highly revered in our culture, have concluded and are concluding things that our mainstream culture treats as heresy. Many of them are cited here. My intention, in fact, is, as much as possible, to let these wise and able explorers present their points of view, and only to draw what they have to say together into some coherent tapestry, a tapestry that I believe is strikingly persuasive, a tapestry that says reality may not be exactly what we've been told it is.

I echo the sentiments of Holmes Rolston, who in his book, *The Three Big Bangs*,[2] observes this: "No one can know firsthand the details of all we survey; that would require a mastermind in cosmology, microphysics, evolutionary history, paleontology, genetics, molecular biology, neuroscience, psychology, logic, and ethics. Still, I write with the persuasion that good philosophers and theologians, good inquiring minds, can look over the shoulders of those doing these things, and spot what is metaphysically interesting. In one sense, doing this, we are Jacks of all trades, masters of none. In another sense, we are more inclusive, more comprehensive than the scientists" (Rolston, 2010, p. xi).[3] Stepping back from the specific findings of scientists, researchers, and others to view their conclusions holistically, we will see, leads to some surprisingly consistent and persuasive, yet unconventional findings.

Over the decades of my life, I found myself generally curious about much of what our culture has come to know, speculate about, and declare to be true. I did a lot of wondering. Then, one day, I guess exploring those things just became important for me to write about. My primary intention now is to get as clear as I can about my best sense of what the universe is and means, and what I can intuit about my and our purpose here. I've discovered some things, and reached some conclusions I think other folks might be interested in hearing about. I studied what science says about cosmology and astrophysics, for instance, and about cause and effect and the challenges of quantum theory,[4] which, I learned, underlies all of what we experience on the

---

2   In this introduction, I cite quotations that are not referenced subsequently. If a quotation is not cited here, that is because it will be when referenced later in its fuller context. Also, a number of quotations appear several times, in varied contexts. I regard these as key ideas, which the reader can treat similarly to the unifying, repeated chorus of a song.

3   Rolston is an esteemed environmental philosopher emeritus, theologian, and author of numerous books. He possesses an undergraduate degree in physics and mathematics, has a master's degree in philosophy of science, and is also a naturalist, according to *Encyclopedia Britannica*.

4   *Quantum theory* is the primary way that modern physics explains matter and energy on the atomic and subatomic level. It will serve as the basis as well for much of what I am articulating in this book.

material plane. And I introduced myself to a number of religions, and to gurus, saints, mystics, and sages, to see what they knew and what they'd concluded about reality.

I saw that there were (surprise!) a colossal number of ways to understand reality, to explain it, and to live life. I came to realize that in any religion, political philosophy, or given social organization or movement, there is a prescribed and "proper" way to think and do and be.

I found that to be true in the discipline of science as well. I saw that in science, there is the "standard model," an agreed-on way to understand what the data we've assembled from scientific study and research mean, an agreed-on interpretation of the data. However, there are lots of assumptions, holes, biases, discrepancies, contradictions, and unknowns in that model, it turns out, and *that* means that much of what we hold as true and dogmatic in the standard model in science is significantly premised on *beliefs* rather than proven *facts;* on what is conjectured rather than what is determined; and on assumptions that are preferred or are comfortably consistent with the model rather than on the raw data we now have available to us.

There are some conspicuous preconceptions that underlie the standard model, which results in the standard model itself beginning to look an awful lot like the credo of a religion. It turns out that, when we speak of the collective body of scientific knowledge, we are expected to faithfully adopt an orthodox vantage point, an established and specific *belief system*, a one-right-way mentality—one that is not necessarily based in verified truths, and one that is either blind to or scrupulously suspicious of any perspective that challenges the conventional point of view.[5]

What are my credentials for talking about such things, and for drawing such conclusions? Whatever my formal background was, it wasn't that of a scientist or mathematician or sage, or even philosopher. The best credentials I personally can muster for the subject area of this book is that I am curious, that I've pursued these curiosities, and that I am a generalist who likes to link what we may know in one discipline with others, and see what shows up. The voice out of which

---

5   A good, though by no means not the only, exploration of the politics of contemporary science that insists on a standard dogmatic approach is well laid out in Rupert Sheldrake's book, *Science Set Free.* Also see *The Biology of Belief,* by Bruce Lipton, who describes some of the issues with the dogmatic assumptions of traditional science from a more personal perspective.

I am writing is that of a layperson who wants to understand what, in a big picture way, we have come to grok[6] as humans about the nature of reality. And I want to share this with others.

Of all the things the hippie era introduced into our society, one thing it brought into our awareness was an opposition to the Vietnam War, which planted a seed in me about war in general—that killing other humans in any war or in any other way makes no sense at all. Killing another human for me was equivalent to killing a part of myself, and so when Uncle Sam came calling, I applied for and received conscientious objector status. That required me to serve two years in a way that would support my country, but nonviolently. I ended up serving two years in a psychiatric hospital in a poorer area of Chicago, which put me on a path to eventually becoming a clinical social worker, and for 34 years so far I have worked as a psychotherapist where I had the spectacular opportunity to become intimately involved in the lives of my clients.

Psychotherapy was my life's profession, and I recognized that whatever issues people showed up with, the common goal was to heal and to grow and to evolve. Everyone was on a path toward further growing and becoming, and issues and problems that all of us have could be the means to that deeper growth. I was curious about human development and its further reaches, and I learned about research that examined that question, such as the work of Suzanne Cook-Greuter, and Clare Graves and the Spiral Dynamics research, which we will talk about in Chapter Ten. Our values evolve with experience and our changing life conditions, but they do so predictably and sequentially, just as a child learns to crawl before walking and before running. Those few that have the opportunities and mindfully seek them out may find themselves at the further reaches, the place where one's spiritual perspective guides our lives and our choices. My curiosity led me to learn about spiritual leaders—Thich Nhat Hahn, Catherine of Sienna, John of the Cross, Jesus, Buddha, Mohammed, Deepak Chopra, Wayne Dyer, A.H. Almaas,

---

6   According to Wikipedia: Grok /ˈgrɒk/ is a word coined by American writer Robert A. Heinlein for his 1961 science fiction novel *Stranger in a Strange Land*...The *Oxford English Dictionary* summarizes the meaning of *grok* as "to understand intuitively or by empathy, to establish rapport with" and "to empathize or communicate sympathetically (with); also to experience enjoyment."

Aurobindo, Thomas Aquinas, Francis of Assisi, Ram Dass, Meister Eckhart, Ken Wilber, Barbara Marx-Hubbard and others. Out of high school, I studied to be a catholic priest for two years—on my journey to becoming an agnostic, then an atheist, then an agnostic again, before arriving at what I hold to be true now. I began reading the works of numerous scientists, a practice that continues to today. Quite a few of those writings are referenced here. Then, beginning in 1995, I began meditating for an hour a day, opening myself again to consider the possibility that there could be life and intelligence at the heart of the universe. Now, I believe that there is a life force that pervades the universe—that *is* a universe which connects us all—that is conscious, deliberate and intentional, emerging and evolving, intelligent, compassionate, relational, and immensely creative.

In my psychotherapy practice, a few key notions became kind of guidelines. As my work more and more was with adults and couples, one's sense of self, personal story, and quality of relationship with self, a partner, others in the world, and the world itself were the themes that seemed most helpful to keep in mind. In 2014, I wrote a book, *Co-Creating a Brilliant Relationship: A Journey of Deepening Connection, Meaning, and Joy,* that brought these ideas together. The last chapter, "Co-Creating a Brilliant Relationship with All That Is," turned out to be the seeds that over time germinated into this book.

Yet the authority and the wisdom behind the subject of this book is, in the end, an extraordinarily prestigious collection of expert, wise, seasoned, illustrious, and cutting-edge scientists, philosophers, teachers, and others, whose thoughts and writings are liberally referenced here. Together, they persuasively argue that there are other points of view that answer questions the "standard model of science" cannot.

We'll start with an overview of our universe from the perspectives of astrophysics and cosmology. Then we look at the infinitesimally minuscule plane of existence and see what traditional physics and quantum science say about atomic and subatomic reality. Connections among diverse research, experiments, and explorations in an array of the fields of science along with inferences and conclusions expressed by a range of researchers, theorists, and other well-known and highly regarded scientific experts, we'll find, point us, inescapably, to several stunning, exciting, and paradigm-altering possibilities, which we're going to describe and explore in these pages.

Next, we'll look at notions such as consciousness and life—essential, yet, perplexing and poorly understood aspects of the reality we live. If we were only to acknowledge that, because we are conscious, consciousness itself may be the medium of and the key to our understanding our reality—the basis of our reality—then we perhaps might be open to considering other thought-provoking notions we'll present, such as the idea that what is real may be a virtual reality as opposed to a literal, substantial, physical one. That, of course, is a notion that would challenge all we have been taught about what is real in our culture and in our world, but it is one that resonates surprisingly with quantum theory, and it offers several insightful intuitions about reality we may never have considered before.

How far down does consciousness of some kind go? Humanity only? Certain "higher" animals? Plants? One-celled animals? Rocks? Atoms? Electrons? If consciousness is primary, does it exist in some form in all these "material entities?" Michael Talbot, in *The Holographic Universe* says, "In a universe in which all things are infinitely interconnected, all consciousnesses are also interconnected." (Talbot, 1991, p. 60). We'll take a close look at these questions and other similarly challenging questions.

How do we actually define what life is? We'll see there is no common agreement about that in science. What *is* living? How did life emerge, and from what? Further, what is intelligence and how does it show itself in our world? These are important questions we'll examine, and perhaps surprisingly, we will see that the answers we uncover are quite inconsistent with the message of the determinist, materialistic, standard model of science.

We'll also examine how we believe life and consciousness (mind) "began," and we'll explore the concept of evolution and what we've come to know—and what we've come to believe—about its unfolding. We'll look at animal intelligence and plant intelligence, and at some of the revealing, befuddling, and paradigm-challenging new data researchers are arriving at.

The ways life and intelligence have developed and evolved in our world will point us, we'll see, to the centrality of consciousness in how we understand "reality." Incredibly, we will discover that evolution itself is an intentional and deliberate, intelligent, and conscious process. It is the universe unfolding. We may even find ourselves considering the conclusion that the universe—that is, all that exists—is living, and has, on some level, the "capacity to know!"

Understanding our world in this way—an unabashedly contrary conception—allows us to much more insightfully comprehend the underlying quantum reality as well, and *its* essential intelligent process. We'll suggest a perspective that offers us a sense of a deep, sacred purpose for our existence in this reality, one that invites us to wonder about the spiritual (though not religious or dogma-bound) relationship we might discover between our own unique, personal, intimate sense of self and this transcendent, numinous universe, of which we are a part.

Quantum theory, and relativity theory[7] have each been comprehensively studied and validated as true, as the way the universe "works," and there is meager challenge to these findings in any corner of the discipline of science. Each unequivocally conclude that, ultimately, there is only subjective perception—a topic we'll speak about in detail as we proceed. Since this is so, *it is the perspective one takes* which determines (for any of us) what we perceive as "real."

My hope is that what is written here contributes to a deepening and expansion of the reader's perspective of that which our glorious universe consists, and so suggests a deeper meaning and purpose for our existence here.

---

7   Quantum theory claims a number of truths that contradict our physical reality assumptions. Subatomic particles behave in ways that are mind boggling—starting with the premise that *there is in reality nothing that is physical.*

The premise of Einstein's General Theory of Relativity is that space is curved, so that the presence of mass/energy determines the geometry of space, and the geometry of space determines the motion of mass/energy. A significant aspect of his theory states that space-time is relative, and so unique to each observer.

The holographic principle suggests that the contents of the universe originate from a boundary surrounding the entire cosmos. As difficult as this idea is to conceive, it is now thought to have mathematical and scientific validity. A hologram is a two-dimensional or flat surface that appears to have a third dimension, giving it an illusion of having depth, when in reality it does not. We'll come back to these concepts in our further discussion.

PREFACE

A simple way to frame what we are considering here is to think in terms of perspectives. In our conversation, we'll be thinking in terms of three particular perspectives, which we'll often reference, each of which might be considered "real" from its particular point of view. The most immediately accessible and absorbing perspective is that of the material world, which we perceive in a physical and sensory manner: *the Materialistic Paradigm:* what is physical is all that is real, and everything is mechanical, cause and effect.[8] Despite how this perception can appear to be so persuasively real, we will unarguably see—given the data from quantum physics—it simply isn't, and it can't be. It is a story, an illusion. Nevertheless, because we experience the physical world as real, we are inclined to interpret and understand our universe as if it indeed was physical.

In the second perspective, *the Consciousness Paradigm,* we participate as choosers with the universe, and we recognize that what emerges is not really physical at all, but virtual, merely a simulation! So we consider, mindfully, that this story we are immersed in is imaginary. It isn't occurring anywhere but in our mind. What we see, experience, and live is an "as if" reality, a virtual world that we consciously and subjectively experience, actively participate in, and make choices about. We intend, we choose, and we act, deliberately. From this perspective, we understand the world as something very different than as the physical cause

---

8   Although a mechanical, deterministic, and material-only universe means that there is no choice, all of us act on the premise that we are choosing all of the time. So which is true, as both cannot be? If there is choice, a purely material-mechanical (cause and effect) universe must be a false understanding of our existence. If there is no choice, than we are mindless automatons. More on all of this ahead!

and effect world seen from the material only point of view.

So what's real beyond the virtual game? That's the third perspective of reality, the *Quantum Mind Paradigm*. From this point of view the entire universe is one interconnected, living, intelligent, deliberate, and evolving entity. From this view, time, space, physicality, choice, and the world itself are unreal—only reflections of something beyond what we can experience.

Each paradigm or perspective is "true" from a certain viewpoint, and each offers insights about the universe of which we are a part. How we understand our being here depends on which of these three perspectives we operate from.

By way of a further introduction, what follows is a brief overview of the book, chapter by chapter:

Chapters 1-4 introduce quantum theory and its implications. Chapter 5 acquaints us with the notion of consciousness and how it might play in our universe, and Chapter 6 examines the Consciousness Paradigm, exploring the point of view that our world may ultimately be a virtual reality, not a physical one. Chapter 7 investigates our understanding of life itself and its origins, which date to the birth of the universe. Chapter 8 surveys life on Earth and documents the pervasiveness of consciousness, intelligence, and agency, in glaring contrast to conclusions of mechanistic science. Chapter 9 looks at the dynamics of evolution, arguing that it is a living, intelligent, selective process, not an inert, mechanistic, mindless one. Chapter 10 describes the evolutionary process as moving from purely biological to a process that is also, with the appearance of humanity, noetic—mental, conscious, knowing—resulting in an evolutionary dynamic that speeds up emergence and growth, and invites humanity to partner with the universe to co-create what will be next. Finally, Chapter 11 delves into the Quantum Mind Paradigm and articulates the point of view that Consciousness is all that is. "We are products of mind: conscious, emergent, choosing, living, and co-creating, in a vast ocean of interconnected presence," evolving as one to richer and deeper levels of being.

Chapter One, *Universal Beginnings—The Big Bang*, describes, in the terms of elite scientists, a stunning, vast, and enigmatic place, where the term "clockwork precision" is a mere faint and insufficient description. Right off the bat, we hear scientists, from the founders and developers of quantum theory to the sophisticated elders of present day physics, compelled to use, as we'll shortly see, surprisingly expressive and passionate, and even non-scientific terms—such as "superagency," "superintellect," "universal mind," "organizing principal," "miraculous," "divine," "purpose," and "existence of a conscious and intelligent mind," among others—to refer to our universe. In our present-day rush to flee anything that might sound like a first cause, theistic or otherwise, or a reality that is other than materially-inspired, and in our determination to relegate any spiritual striving to magic and myth, there is a strong impulse to reject such expressions as child-like, merely metaphoric, or fanciful.

This materialist doctrine which assumes that everything is physical and all that we perceive in our world has a purely and solely physical origin, is the tightly held bias of our contemporary cultural credo, at least in the first world. Yet, hearing the wide-ranging, awe-filled, and incredulous expressions of these sage and diverse elders, perhaps, may invite more of an openness to wonder, rather than a knee-jerk reaction of out-of-hand rejection.

Chapter Two, *Quantum Reality—The Underlying Structure of the Cosmos,* introduces quantum physics, the study of the micro arena of our world. What is big is built on what is small, so having an overview of how quantum physics plays out in our world is essential. We survey the quantum realm of the atomic and subatomic, where the familiar rules for the world in which we live become suspended, and where in its place the perplexing dynamics of the quantum domain, (the micro world out of which the macro world is built), show themselves to be in stark and incomprehensible contrast to our everyday reality. Importantly, quantum physics says that, in the subatomic world (out of which everything "physical" is constructed) *there is nothing physical that exists.*

We will also briefly chronicle the ways western humanity has understood reality in various historical periods.

Chapter Three, *Quantum—The Strange Features of Quantum Dynamics,* describes this unfamiliar world that quantum physics has proven to be what is, and the rest of the book is premised on what is described in Chapters Two and Three, so they're important.

This chapter exposes a world where time and space are in the eye of the beholder— there is *not* an objective world we can agree on—and indeed there is no matter at all; subatomic particles are waves and waves are particles which zoom in and out of existence; they are interconnected across infinite space; and it is a world wherein we cannot affirm that anything exists without conscious observation. This is the underlying reality of existence—not according to some hyper-creative science fiction writer, but according to the data from the discipline of physics, and demonstrated by the compelling outcomes of our quantum techno-miracles, and by the visible, concrete results of its implementation in our world economy. We must reckon with that enigma: that the everyday sensory world in which we live, is, per our best science, non-existent as such! I ask you not to take my word for it, but, rather, take the word of "the best and the brightest" in our culture. We look specifically at the raw experimental data from quantum theory and quantum practice, and examine the specific apparent oddities that conflict so brashly with the "standard model" of orthodox science.

B. Allen Wallace sums it up: "Quantum physics is... (an) odd piece... Here we have mathematical formulas that work like a charm. But the reality they refer to is mind-boggling. Waves act like particles and particles act like waves. Things lack certainty until measurements [i.e., observations] are made, but some vital measurements cannot be made at all. This...implies that systems as a whole are somehow indeterminate. Furthermore, events in one locality seem to be influenced instantly by events *even light years away*" [emphasis added] (Wallace, 2008, p. 115).

Chapter Four, *Quantum Reality—Interpretations and Implications*, is the most technical chapter of the book. We explore a bit the differing ways scientists make sense of the experimental and applied data from quantum mechanics. The rules of the very small are very different and conflict with the rules of our common everyday experience, and it is this glaring contrast of the underlying quantum reality that makes it so difficult to understand or even accept. What are the different ways the data can be interpreted and understood? Here, bizarrely, we find that there *is* *no* agreed on interpretation! The original interpretation, what is called "the Copenhagen interpretation," formulated by the original founders of quantum theory, is still the dominant interpretation, preferred by most physicists. However, there is a lot of competition—there are other interpretations, which I will introduce in a general way.

The commonality among the great majority of ways science has tried to understand quantum physics is the assumption that we live in a material, physical universe, and thus, that matter came first—before consciousness, life, and the nuance of native human consciousness. It is noteworthy that effectively *all* of the frequently considered quantum mechanical interpretation candidates begin with this assumption of the primacy of matter: that the big bang consisted of matter first, and out of matter, as epiphenomena, or secondarily, life, mind, and self-reflective consciousness arose. The primacy of matter is treated as an unarguable and proven fact by mainstream science, *although no specific proof has ever been shown.* Mainstream science prefers to think matter emerged first in the universe because it is consistent with our experience of a physical world which we can examine, and because to conceive of consciousness co-emerging with or preceding matter raises a host of questions science has been unprepared to consider and unable to address. With that assumption of matter first, no generally persuasive interpretation of quantum mechanical data, and no integration or synthesis among traditional physics, quantum mechanics, and relativity theory, has been possible. The Materialistic Paradigm based in a matter first mentality doesn't have the "experience" or "language" to make sense of non-physical particles and waves and their behavior.

Physicist Karl Pribram tells us: "It isn't that the world of appearances is wrong; it isn't that there aren't objects out there, at one level of reality. It's that if you penetrate through and look at the universe with a holographic system, you arrive at a different view, a different reality. And that other reality can explain things that hitherto remained inexplicable scientifically." (Talbot, 1991, p. 11). From the point of view of the Materialistic Paradigm, there *are* objects. From the Consciousness Paradigm viewpoint, a virtual and holographic reality, what we are perceiving as a material reality is seen to be illusory.

Each of the customary interpretations of quantum physics argue for a reality of a physical world we cling to, yet none really resonate or provide us with an intuitive grounding in understanding what's going on. I believe the primary reason for this is that the assumption of matter first is a false one, and I argue that, indeed, consciousness preceded everything, matter included. Therefore, in a sense, this chapter is least important, given that distinction. It's also technically challenging to a degree, though I try to keep it simple. It may be valuable for some readers to glean a sense of how contemporary materialistic science looks at quantum data and continues to try to understand it through a physical lens, but otherwise the chapter might be skipped.

Chapter Five, *Consciousness—The Ultimate Nature of the Cosmos,* focuses on the incontrovertible experience of consciousness itself, and argues that its centrality can help us understand the world we live in in a strikingly different way than physicalist points of view. Consciousness is a predominantly interior reality, something we each experience discretely and uniquely, from within.[9] It can even be argued that there is no universe other than the one inside our minds, with all we experience being a projection outward. Chapter Five tries to understand the world from the vantage point that what we experience as material is not, but merely an illusion, like the desert illusion of the oasis that I am positive is there. Except it's not; it's only an exceptionally convincing deception.

But if our experience is illusory, how does comprehending that change our understanding of our existence? Perhaps consciousness or mind *precedes* the arising of the material world. Physical realism, (matter is primary), dualism, (both matter and mind exist but are independent of each other), idealism, (mind is primary and preexistent), and panpsychism (mind and matter co-arise), are the customary alternative points of view, and we explore the implications of each of these ways of understanding what is real. [10]

Incredibly, we arrive at the conclusion which Max Planck, the first founder of quantum theory, came to: "I regard consciousness as fundamental. I regard matter as derivative from consciousness. We cannot get behind consciousness. Everything that we talk about, everything that we regard as existing, postulates consciousness" (Sullivan, January

---

9   We will talk in Chapter Nine about intersubjective consciousness, which does not reside entirely within, but between us, and does not have any physical manifestation.

10   Christian de Quincey teases these ideas out specifically, as we'll see in this chapter.

25, 1931, "Interviews with"). This is an idealist stance, "quantum realism," which affirms that consciousness precedes the material world. If this is so, we see that we are de facto part of a living universe, *(a universe that perceives must be alive),* and we are in relationship with the universe, as scientists from a variety of disciplines here argue.

For most folks, steeped in the orthodoxy of contemporary materialist, reductionistic, deterministic science as they are, here is where the eyes start to roll and the sarcastic, dismissive, mocking commentary sometimes begins. As strange and unfamiliar as these ideas sound, they are more in alignment with the conclusions of quantum theory—*which science has indisputably declared as the underlying reality*—than any of the inconsistent and insufficient conclusions of a cause and effect material basis (physical realism) could be!

"The Austrian philosopher of science, Ernst Mach (1838-1916), who influenced Einstein's theory of relativity, explicitly rejected a mechanistic conception of matter, and wrote, 'Properly speaking, the world is not composed of "things"… but of colours, tones, pressures, spaces, times, in short, what we ordinarily call individual sensations'" (Sheldrake, 2012, p. 118). As challenging as this sort of assertion is, it reflects the truth of our existence, as we will abundantly affirm in these pages.

We need only alter one premise, and things begin to fall into place: It is not matter that is primary, it is mind. As Erwin Schrödinger, who developed several fundamental principles of quantum theory, (and was famous for the "Schrödinger's Cat" thought experiment), declared: "What we observe as material bodies and forces are nothing but shapes and variations in the structure of space. Particles are just *schaumkommen* (appearances)" (Schrödinger, 1995). So, again, per quantum theory, matter is merely an appearance of mind, insubstantial in itself!

Chapter Six, *An Alternative Conjecture—The Universe is Virtual,* explores this weighty premise: that there, indeed, is nothing physical; that what we are living is virtual, as though we were characters on a holodeck. Because there is nothing physical, however, does not mean there is no life. In fact, without physicality, all that there could be is our consciousness, our perspective, our lived, intelligent experience of the universe unfolding. All there could be is action, movement, and mind.

This chapter proposes that our world is, in ways, analogous to a virtual computer game. We engage in the game, and for a time we are singularly focused on it, as if it were our entire reality. We give identities to others—avatars in the game—as well as to ourselves. We are conscious of the game, and play it intelligently, yet, in the end, we realize it is only a dalliance in fantasy. In the end, we realize that reality is something far more intricate and meaningful than the game we've been playing.

We consider here this notion that consciousness is primary. The current scientific understanding of the beginnings of our universe (the big bang and inflationary theories) says that it is the *quantum fluctuations* that were present in the *infinitesimal vacuum* which spawned the big bang: it is a *quantum presence* which germinated physical manifestation. If the "quantum fluctuations in the infinitesimal vacuum"[11] are not physical, how then could matter be primary?

So here we begin to connect the dots: quantum states and mind are one and the same thing: "in their essential nature, quantum fluctuations are the stuff of consciousness and will," says physicist Evan Harris Walker.[12] And he goes on to say, "We discover that in the beginning, there was the Quantum Mind, a first cause, itself time-independent and nonlocal that created space-time and matter/energy" (Walker, 2000, p. 326). Essentially, quanta, subatomic "packets of action," are not material; *rather, they are the stuff of aspects of non-material consciousness.* Walker, particularly, points to this link between mental and quantum processes, originating as "Quantum Mind." Neither quanta nor consciousness are physical, existing only in a noetic, mental sense.

Though Walker is unambiguously persuaded, the link between quantum fluctuations and consciousness is, of course, unproven. The concept of "quantum fluctuations [being] the stuff of consciousness

---

11   "'Where did the universe come from?' writes physicist Evan Harris Walker, in *The Physics of Consciousness*. "How could everything appear from nothing?'…In the last few years physicists have developed a new theory of the Big Bang's first moments, a history of the universe as it existed between the age 10^{-43} second and 10^{-35} second. The theory, first developed by [Alan] Guth, is called the inflationary theory of the universe. It describes how random quantum fluctuations in an infinitesimal vacuum would have led to an incredible expansion…" (Walker, 2000, p. 323).

12   Evan Harris Walker is considered to be one of the founders of the modern science of consciousness research. First to advance a theory that the nature of consciousness is tied to physical principles of quantum mechanics and based on quantitative physical and neurophysiological data, he has made significant contributions to the measurement problem in quantum mechanics and originated (as described in this work) the "Quantum Observer Theory" of state vector collapse. He died in 2006.

and will" leads to a conjecture that, just as an avatar in a virtual game has no physical reality, but is virtual, so too may be our universe. In fact, if there is nothing truly physical, as we have *already* found, then we *must* conclude that the universe is virtual. "Virtual reality" means "near-reality:"

"Our experience is an amalgamation of sensory information and our brain's sense-making mechanisms for that information. If you provide your senses with made-up information, your perception of reality would also change in response to it," as the Virtual Reality Society website explains. "You would be presented with a version of reality that isn't really there, but from your perspective it would be perceived as real" (Retrieved from the Virtual Reality Society, 2017, "What is").

Further, *if the world we experience is virtual,* then *de facto what is real is mental—conscious, and thus alive*—and not at all physical!

And that is what quantum realism, that is, reality that is non-physical, or virtual reality, contends: what we experience through our senses are symbols of an underlying reality, perceived through our conscious observation, but not the underlying reality itself. Within a virtual game, one's experience is real, but applies only within the game: it is not the total and complete reality, which would include the player, the programmer, and any other attributes of those broader realities. A TV show seems real, but it's not. Our experience *seems* real, *seems* physical, but it's not. Chapter Six considers these ideas, argues for their likely validity, and explores their implications. Ultimately, the virtual reality conjecture, with its implicit premise of consciousness/mind as primary, offers an interpretation of the quantum data that is consistent and sensible: matter is an epiphenomenon of consciousness, quantum mind.

Chapter Seven, *Life—The Essential Process of the Cosmos* carries us beyond physics and matter to chemistry and biology, and questions of what life is and how it appeared, and we find that, again, if we understand the universe as an intelligent, deliberate, and an evolving, albeit virtual, event, the concept of life gets cast more richly, deeply, and holistically as no less than *the essential process of the cosmos.* We will see that organic molecules existed in the *first, earliest* actions, gases, and chemistry of the embryonic big bang, and are abundantly present throughout the universe—indicating that the capacity for life was present in the universe "from the beginning:" a life force, consciousness itself, we will see, pervades the universe, is the universe. We'll look at life's beginnings on Earth, and what science knows and conjectures about its beginnings here. We will see that life expresses itself with ever deepening complexity, that the process of its emergence is holistic, and that it involves the entire ecosystem of the Earth and even the universe.

Then, in Chapter Eight, *The Capacity to Know—Intelligence in Nature*, we explore how the life force looks on our planet, and we immediately see that the Earth and the universe involve intelligence, or a capacity to know, throughout, flowering out from the organic molecules distributed throughout the early universe, to the "non-living" materials of our planet, to single-celled beings, to flora and fauna, and to humanity. We also see that what we call life on Earth does not consist of discrete beings, but truly consists of one interconnected being, conscious throughout. We will survey a variety of the animal and plant life that inhabit our planet, and we'll find that each species demonstrates forethought, cognition, decision-making, interaction, and communication with same and different species, a pervasive capacity to know, an ability to learn and adapt, and an inherent experience or consciousness. Some exhibit tool use, individuality, theory of mind, empathy, and even personality. Everywhere, the Earth's ecosystem displays awareness, intelligence, and agency. Even what we view as non-living aspects of our world have been parts of stars and rocks and air and past living beings. Each atom within us has cycled between things we think of as not living and things we think of as alive. The great California Redwood is ninety-nine percent dead wood—yet we consider it as one living whole.

Next, our conversation moves to Chapter Nine, *Evolution—Emergence, Agency, and Innovation*. Of course, in the theories of Darwinian and neo-Darwinian evolution, such teeming agency is invisible, as we'll see. Natural selection and random genetic mutation, arbitrarily, blindly, and capriciously are said to morph dead molecules and inert chemicals into life, mind, consciousness and native human consciousness, and eventually such facets of our world like beauty, truth, and goodness.

Stuart Kauffman writes: "For the reductionist, only particles in motion are ontologically[13] real entities. Everything else is to be explained by different complexities of particles in motion, hence are not real in their own ontological right. But organisms, whose evolution of structures and processes, such as the human heart, cannot be deduced from physics, have causal powers of their own, and therefore are real *emergent* entities in the universe. So, too, are the biosphere, the human economy, human culture, human action" (Kauffman, 2008, p. 3).[14] Kauffman articulates here the basic disparity between a mechanistic model and the consciousness (agentic/emergent) model we are elaborating.

The traditional standard model of mechanistic evolution asks one to ascribe to a miracle far more incredulous than water into wine: that the lifeless and soulless become the animated and engaged. If matter must be primary, this is the miracle required. A far more, if you will, organic, possibility is that mind was always present, and from its brilliance, mortal and material being intelligently and creatively emerged. Additionally, agency, the capacity to choose, emerged, displaying itself in pond lilies, gnats and gnus, and us.

We will see that the process of evolution—in the universe and on our planet—is a deliberate, intelligent, choice-making, and emergent process, not at all mechanical and random, as traditional deterministic understandings would hold.

We conclude this chapter by pointing again to the notion that, indeed, *we are a part of a living cosmos,* a sentiment expressed succinctly and eloquently by the eminent biologist Elisabet Sahtouris,[15] who says: "We can construct a new scientific model...that takes into account the entire gamut of human experience and recognizes the cosmos as fundamentally conscious and alive ... a holarchic, evolving, intelligent, process intrinsic to the cosmos itself" (Sahtouris, 2007, Chapter 7). Sahtouris' frame of this "new scientific model" to counter the "standard model" of science is actually far more tenable as we will argue throughout this book, and it represents a holistic, exciting, vital, harmonious,

---

13   Merriam Webster dictionary: "relating to or based upon being or existence."

14   Kauffman is a researcher in complex systems and a theoretical biologist. He is known for his views that biological complexity of systems and organisms may be a product of self-organization as much as from natural selection.

15   Sahtouris is an internationally known evolutionary biologist, futurist, and author of numerous books, including *EarthDance: Living Systems in Evolution.*

and creative option to a view of our universe as a machine.

Chapter Ten, *Native Human Consciousness—The Exponential Evolution of Consciousness* considers the notion of the evolution of consciousness directly. We examine what we might call the developmental levels of consciousness—from the existential awareness of a rock to human self-reflective knowing. Harville Hendrix reflects, for example, that he "finally adopted the position that all things, organic and inorganic, are conscious because they are made of Consciousness, but not all things are alive. The universe, including us, is conscious, not as a quality but as essence—and that part of the universe that is us (biological beings) is conscious and alive. Thus rocks, which clearly are not alive, are nevertheless conscious" (de Quincey, 2005, pp. xix-xx).[16] This is a challenging and controversial topic which we'll explore here.

Several other scientists and philosophers, which we'll specifically reference, (de Quincey, Freeman Dyson, and Graham Harvey, to name a few), would go even further—even rocks and particles possess a degree of existential awareness, and thus participate as living features of the universe. Physicist David Bohm's view too is that: "Even a rock is in some way alive...for life and intelligence are present not only in all of matter, but in 'energy,' 'space,' 'time,' 'the fabric of the universe'" (Talbot, 1991, p. 50).

Certainly, since everything is one interconnected being, as scientists and philosophers and theists contend, and as quantum dynamics asserts, then the totality must be conscious and thus, alive. Further, we again examine the context that is the universe and the ecosystem that enfolds the Earth, appreciating the universality of life, mind, and agency.[17] Consciousness witnesses a world that is seemingly external to our individual beings, and it also discerns the world within, internal consciousness. (Yet we will take note here that, at its foundation, our experience of what is real appears, finally, to exist only within)!

Our human experience, our native human consciousness, has carried us further than any other entity we have discovered. Humanity has the capacity both to know and *to know that we know!* We'll explore a

---

16   Christian de Quincey is a professor of Philosophy and Consciousness Studies at John F. Kennedy University and former editor of the *IONS Review* and *Shift*, the magazine published by the Institute of Noetic Sciences in California.

17   Capacity for choice driven action.

bit how that exponentially expands what we as a species are capable of, and we'll come to understand that we, alone—as far as we know—are capable of partnering with the universe to deliberately author what might emerge next. We have transcended biological evolution with our capacities of mind and are thus able to co-create with the universe, and to evolve beyond the capacities of our biological roots.

In addition, humanity has manifested a capability to share our individual conscious awareness with each other, resulting in deliberate invention and creation, and in a shared intersubjective consciousness that exists within each human and beyond each human in a communal, non-physical space: a final display that consciousness is beyond matter. Steve McIntosh declares "These systems of human relationships are not 'in the air,' nor are they merely in our minds. They exist in the intersubjective domain of the internal universe." Consciousness, the tool of our future unfolding, is beyond matter. It is consciousness, and particularly self-reflective consciousness, that bends matter according to its preferences. We are the ones who will say what we do with this native human consciousness, this self-reflective and intersubjective noetic essence that is who we are.

In Chapter Eleven, *The Soul of Our Universe—A Cosmic Presence,* we, finally, bring all we've said together, and reach the inexorable conclusion that all that is is Consciousness alone, and all that we experience is a subjective and personal perception of our "reality"—not physical, but ultimately solely existing in mind.

Rather than being physical, we come to understand that we are movement and action and events, not things, and we are products of mind: conscious, emergent, choosing, living, and co-creating, in a vast ocean of interconnected presence. Ultimately, we are quantum mind, beyond time and space and matter and energy. We are an aspect of the first cause of our existence—the booter-uper—and we are interconnected and the same. We are each conscious, and, like a mirror ball with its tiny individual mirror panels—both different as well as the same—we are each an aspect of One Consciousness.

We grow and evolve, and we may deepen our conscious awareness, which might lead us, as we appreciate our oneness to a greater and greater degree, to a more mindful and intentional focus on care and compassion for ourselves, humanity, and the ecosystems of our Earth and our universe.

In our personal growth, we may reach a point of holding what John Stewart calls "an evolutionary awareness," where we act as representatives of the grander goal of "evolutionary success of future generations." Stewart points out to us that "it is an illusion to see ourselves or other individuals as distinct and separate entities. Individuals are inextricably part of an ongoing evolutionary process…Evolution itself evolves, and living processes get smarter at evolving." Similarly, McIntosh emphasizes that "As our worldview grows wider and deeper, it becomes increasingly evident that the story of evolution in nature, self, and culture has unmistakable spiritual implications." Indeed, we will consider the evolutionary process of the universe itself as a spiritual process, and a primary and meaningful way we engage in spiritual practice.

We come to appreciate the essential point of our being, and, as well, this book. I point out that the "Cosmic Presence is all that is, and it exists within the game and beyond the game. And the game itself exists so we can notice, as a vehicle for us to begin to grok and begin to ever more mindfully participate in this Cosmic Presence."

In this regard, I write in this chapter, "That which the Cosmic Presence, Spirit, is bidding us to do is something we discover personally, subjectively, and uniquely, but the path to discovery gets clearer the more we move up the developmental values-spiral toward greater inclusivity."

Now you know where this conversation is going. The process of moving from an outdated physical and material way of understanding our existence to richer and deeper insights and understandings about what our being here and our purpose here are about is a riveting one, an incredulous one, and one that, I believe will lead you, Dear Reader, toward a mindful, co-creative, joyful, and intentional life in a universe filled with magical things.

PART ONE:

THE GAME

"It surprises me how disinterested we are today about things like physics, space, the universe, and philosophy of our existence, our purpose, and our final destination. It's a crazy world out there. Be curious."

—Stephen Hawking.[18]

---

18 Retrieved from the website http://www.educatinghumanity.com/2013/01/Stephen-Hawking-Quotes.html).

Chapter One:

━━━ ✍ ━━━

UNIVERSAL BEGINNINGS— THE BIG BANG

THE FIELD IS THE ONLY REALITY

"There are only two ways to live your life. One is as though nothing is a miracle. The other is as though everything is a miracle."

—Albert Einstein (Hinshaw, 2006).

As we will powerfully contend in the pages ahead, the implications in the data from a variety of disciplines in contemporary science oblige us to consider an understanding of our world that is astonishingly different from what traditional materialistic science would have us conclude, and from what our culture accepts as true and has taught us to believe to be true. The paradigm-shattering impact of what we are learning, we will see, is clear and unarguable. And transformative.

Bruce Lipton, a biologist, writes, "Ironically, some of the new insights offered by science are so far outside of what we have accepted as conventional wisdom that science itself is having a hard time coming to grips with the implications" (Lipton & Bhaerman, 2009, pp. xx-xxi).

In fact, "for a number of decades respected scientists in a variety of disciplines all over the world have been carrying out well-designed experiments whose results fly in the face of current biology and physics..." declares Lynne McTaggart, (McTaggart, 2008, xxiii-xxiv), author of *The Field, The Quest for the Secret Force of the Universe.* "What they have discovered is nothing less than astonishing. At our most elemental, we are not a chemical reaction, but an energetic charge. Human beings and all living things are a coalescence of energy in a field of energy connected to every other thing in the world. This energy field is the central engine of our being and our consciousness...."

McTaggart describes a remarkable reality, the unity of our universe, stating ..."There is no 'me' and 'not me' duality to our bodies in relation to the universe, but one underlying energy field... 'The field,' as Einstein once succinctly put it, 'is the only reality'" (McTaggart, 2008, pp. xxiii-xxiv). In the pages ahead we will see how the current research and data from a variety of fields of science unequivocally ratify this conclusion, and try to make sense of the implications.

THE BEGINNINGS OF OUR UNIVERSE

The common, most broadly accepted cosmological storyline of the universe, within which we mark the beginning of time and space— our existence—starts in the moment scientists refer to as the big bang.[19] "At some point in the past...the universe was squashed into a single point with zero size," physicist Stephen Hawking tells us. "At that time, the density of the universe and the curvature of space-time would have been infinite. It is a time that we call the big bang" (Hawking, 2005, p. 68). Further, he says, "the old idea of an essentially unchanging universe that could have existed forever, and could continue to exist forever, was replaced by the notion of a dynamic, expanding universe that

---

19   "The Big Bang Model is a broadly accepted theory for the origin and evolution of our universe. It postulates that 12 to 14 billion years ago, the portion of the universe we can see today was only a few millimeters across. It has since expanded from this hot dense state into the vast and much cooler cosmos we currently inhabit," states NASA's Wilkinson Microwave Anisotropy Probe report. "The Big Bang did not occur at a single point in space as an 'explosion.' It is better thought of as the simultaneous appearance of space everywhere in the universe. That region of space that is within our present horizon (the limiting distance from which we can have received information since the Big Bang, 13.7 billion years ago, ... today... close to 45 billion light-years), was indeed no bigger than a point in the past...." (Retrieved from www.wmap.gsfc.nasa.gov).

seemed to have begun a finite time ago, (around 13.7 billion years), and which might end at a finite time in the future" (Hawking, 2005, p. 49).

Whatever the phenomenon of the big bang was, what we experience as the world[20] is the result. You and I and all that we know to exist were held, somehow, as a creative possibility within the big bang. I and we flowed from the big bang—and all that is in the universe we know flowed from it—*our very existence here emerged out of the "stuff" of the big bang.*

Whatever that "stuff" was, it contained the ingredients for our eventual becoming. Holmes Rolston noted, "...one of the surprises of contemporary physics is that the human person is composed of stardust, fossil stardust! ...we humans are leftover nuclear waste" (Rolston, 2010, p. 32).

Nuclear waste in the form of humans: "At some point in the far distant past," the National Geographic series *The Known Universe* remarks, "we were dealt a pretty good hand when an ancient star died a spectacular death, and seeded our little corner of the galaxy with everything we'd ever need: all the raw materials—iron, carbon, calcium, oxygen and other elements—that would billions of years later lead to this: life. Our bodies are composed of star stuff, the debris of exploding stars that manufactured the atoms of which we're composed. Every atom in every one of us was at some point part of a star. In fact, everything around us was once part of a star... we're all recycled atoms" (Hickman, *The Elegant Universe,* 2003).

Or, as Jimmy Buffett sang, "It's something more than DNA that tells us who we are—it's mythic, and it's magic. We are of the stars!" (Buffett & MacAnally, 1999, "Oysters").

Whatever existed "before"13.7 billion years ago is, and likely will remain, ever a mystery. Indeed, Hawking defaults: "Questions such as who set up the conditions for the big bang are not questions that science addresses" (Hawking, 2005, p. 69). Nevertheless, our science *does* speculate about the question of those conditions,[21] but without any

---

20   Except when explicitly quoting others, we are using the words "world," "cosmos" and "universe" as synonymous for "all that exists." The terms "metaverse" and "multiverse" are not equivalent, though, and we'll describe how that is so ahead.

21   For example, such speculation is central to Alan Guth's Inflationary Theory—more later.

accessible data to draw from, these conjectures veer into the jurisdiction of philosophy, not science per se.

"What is not considered a mystery or a question," says Ervin László,[22] "in the sense that it has now been described with some certainty using the best, most recent, and most reliable results of our continuing scientific inquiry, is that about 13.7 billion years ago, (or perhaps as long as 15.8 billion years ago), there was a massive inflation or expansion" (László, 2007, p. 82).

What was it that expanded? We call it a "singularity," but we have no idea what a singularity[23] is. A ball of matter-energy compacted to zero size and bearing all the weight of the sum total of what exists in our universe—burst forth. This expansion blew rapidly outward, first as a potpourri of basic atoms, molecules, and chemicals, and then, quickly, as a main course of more and more complex chemicals and forces. Life and mind—it *had to be life and mind too,* somehow; latent, or not yet noted, or in some kind of potential—as well, were a part of this incomprehensible manifestation. "Over time," Hawking believes, "as the universe has undergone a complex evolution (its) makeup has also evolved. It is this evolution that has made it possible for planets such as the earth, and beings such as we [sic] to exist" (Hawking, 2005, p. 82). If we fast-forward to today, we can see that this universe continues to expand outward with increasing speed, which, as Arthur Eddington, the esteemed astrophysicist described, is like a balloon being blown up.

According to the standard cosmology, before the big bang—before time and space as far as we can document—was nothing. Nothing but *quantum fluctuations* (Hawking, 2005, p. 70). Zero dimensions, then four, (at least). Out of a single point of zero space, a universe emerged—a universe containing matter/energy, life, and mind. This universe continues to expand and evolve, increasing in size and complexity, and containing within it life, mind, and self-reflective awareness—in the form of humans.

---

22 Hungarian philosopher of science, systems and integral theorist, and two time nominee of the Nobel Peace Prize.

23 Only that it is a "place" where mathematical results are infinite, and so they are meaningless.

An Incredible Vastness

"I find it as difficult to understand a scientist who does not acknowledge the presence of a superior rationality behind the existence of the universe as it is to comprehend a theologian who would deny the advances of science."

— Wernher Von Braun, German-American
rocket scientist (Willhite, 2007, p. 89).

We know that we exist on a planet we call Earth, which, along with eight (plus) other planets, all circle our sun. Our sun is a star, like most of the objects we see in the night sky, and we have learned that it swirls around in a tightly knit spiral along with one hundred billion other stars we call the Milky Way Galaxy. We have relatively recently learned, with the launching of the Hubble telescope in 1990, that there are *within the range of our observation* 100 billion other galaxies as well, each containing 100 billion (or more!) stars, other suns. [24] [25] With the naked eye, we can see about five thousand stars, only about .0001 of all the stars just within our own galaxy.

---

24   M.I.T. cosmologist Guth's Inflationary Theory of the universe describes how random quantum fluctuations in an infinitesimal vacuum would have led to an incredible expansion.

25   "70,000 million million million stars—that's the total number of stars in the known Universe, according to a study by Australian astronomers. The figure—7 followed by 22 zeros [or, technically, 70 sextillion]—was calculated by a team of stargazers based at the Australian National University. Speaking at the General Assembly of the International Astronomical Union meeting in Sydney, Dr. Simon Driver said the actual number of stars could be infinite." (As reported by CNN July 23rd, 2003).

Then, quite recently, in December 2010, a study reported in the journal Nature by Yale astronomer Pieter van Dokkum—just took the estimated number of stars in the universe—100,000,000,000,00;0,000,000,000, or 100 sextillion—and tripled it. (As reported by Discover Magazine).   http://blogs.discovermagazine.com/80beats/2010/12/01/the-estimated-number-of-stars-in-the-universe-just-tripled/#.XFt2GVxKiM8

...And how fast things change with new technology: In a *Forbes.com* post entitled "How We Know How Many Galaxies Are In The Universe, Thanks To Hubble," Ethan April 15, 201, Ethan Siegel reported: "...a region of space barely a thousandth of a square degree on the sky — so small it would take *thirty-two million* of them to fill the entire sky — contains a whopping 5,500 galaxies, the most distant of which have had their light traveling towards us for some 13 billion years, or more than 90% the present age of the Universe. Extrapolating this over the entire sky, we find that there are 170 billion galaxies in the observable Universe, and that's just a *lower limit*. Given that there's an astonishing amount of Universe out there that Hubble cannot yet see — and that it can only see the *brightest* of the most distant galaxies — we may yet find closer to 10^{12} (or a *trillion*) galaxies within our visible Universe." . https://www.forbes.com/sites/ethansiegel/2015/04/24/how-we-know-how-many-galaxies-are-in-the-universe-thanks-to-hubble/#41636dc6e514. Interestingly, in 2015, the Hubble telescope laid its "eye" on a planet 729 trillion miles from us, and found, even at such an inconceivable distance, something very familiar to us: water! (Yost, *Invisible Universe Revealed*, 2015)

A Universe Full of Magical Things

The number of stars is inconceivable—if we count *only those we know of*, that number exceeds all the grains of sand on all of the beaches and in all of the deserts on the entire planet Earth! (Hawking, 2005, p. 53). That's not counting other astral bodies such as planets, asteroids, comets, space debris, or other material entities out there.

The immense dimensions of space in which they exist are another order beyond that. "Put three grains of sand inside a vast cathedral," noted Sir James Jeans, "and the cathedral will be more closely packed with sand than space is with stars." [26]

Yet within this titanic magnitude there exists other mysterious aspects to the cosmos: "dark energy," which we believe makes up 68% of the universe and "dark matter," which makes up 27% of the universe (Hickman, *The Elegant Universe*, 2003).

As the universe continued to expand out of the big bang, new stars and new galaxies continued forming, and, also inexplicably, life and conscious mind! Then, in this truly unimaginable largeness, grandness, and mystery, off in some ordinary and minuscule moment in the life of the universe, humanity was born. Cosmologist Brian Swimme (Bridle, "Comprehensive," p. 40), sums it up: "It's really simple. Here's the whole story in one line...You take hydrogen gas, and you leave it alone, and it turns into rosebuds, giraffes, and humans." Science has witnessed the magic, but is silent as to how it could occur.

A VERY PRECISE TINKERING

"A little science estranges a man from God; a lot of science brings him back."

—Sir Francis Bacon.[27]

How the universe we are a part of could come to exist, and how *we* could come to exist, is a question that has stymied humanity all the

---

26 "Our Home in Space." In R.C. Prasad (ed.), *Modern Essays: Studying Language through Literature*, (1987), p. 26.

27 This quotation is commonly attributed to Francis Bacon, whose actual quote can be found in "On Aetheism" in *Essays* (1597): "A little Philosophy inclineth Mans Minde to Atheisme; But depth in Philosophy, bringeth Mens Mindes about to Religion."

way back. The serendipity of so many requisite features of our universe arising concurrently, and the delicate balancing involved in its continued actuality, are realities well beyond our capacity to understand. And yet, this is what is.

Robert Lanza, scientist and theoretician, has concluded, unimaginably, that "The probability of random physical laws and events leading to this point is less than 1 out of 100,000,000,000,000,000,000,000, 000,000,000,000,000,000,000,000,000,000, [or one in one hundred quindecillion]."[28]

John D. Barrow cosmologist, theoretical physicist, and mathematician at the University of Cambridge, observes of the universe: "Many of its most striking features—its vast size and huge age, the loneliness and darkness of space—are all necessary conditions for there to be intelligent observers like ourselves." (Barrow, 2002, p. 113).

Further, in addition to its incomprehensible size and age, and the captivating mysteries of life and mind that have evolved within it, scientists of all varieties remain astonished at our perplexing beginning in what we call the big bang—but they are virtually unanimous that our universe began with this "originating huge explosion," as Rolston calls it (Rolston, 2010, p. 2).

Cosmologist Paul Davies notes the "extraordinary physical coincidences and apparent accidental cooperation" (Davies, 1982, p. 110) of which our universe consists—resulting in a harmonious (and fortuitous!) symphony far beyond any ever heard.

Hawking and Leonard Mlodinow remark as well on the intricacy of this potent yet delicate phenomenon of the birth of the universe in *The Grand Design*. "The emergence of complex structures capable of supporting intelligent observers seems to be very fragile. The laws of nature form a system that is extremely fine-tuned, and very little in physical law can be altered without destroying the possibility of the development of life as we know it. Was it not for a series of startling coincidences in the precise details of the physical law, it seems humans and similar life-forms would never have come into being…What can we make of these coincidences? …Our universe and its laws appear

---

28   Dr. Robert Lanza was selected for the 2014 *Time* 100 list of the hundred most influential people in the world. As quoted on the website http://www.robertlanza.com/why-are-you-here-a-new-theory-may-hold-the-missing-piece/.

to have a design that is both tailor-made to support us, and, if we are to exist, leaves little room for alteration. This is not easily explained, and raises the natural question of why it is that way" (Hawking and Mlodinow, 2010, pp. 161-162). We will see that such a plethora of "coincidences" leaves us with a conclusion that a first cause of some kind is inevitable.

Lanza references the philosopher John Leslie's 1989 book, in which he amplifies on the precise properties we see within the cosmos that have allowed for life to emerge. Writes Lanza, "Philosopher John Leslie… says, 'A man in front of a firing squad of 100 riflemen is going to be pretty surprised if every bullet misses him. Sure he could say to himself, "of course they all missed; that makes perfect sense, otherwise I wouldn't be here to wonder why they all missed." But anyone in his or her right mind is going to want to know how such an unlikely event occurred….'"

Lanza continues: "If the universe is created by life, then no universe that didn't allow for life could possibly exist. This fits very neatly into quantum theory and [esteemed physicist] John Wheeler's *participatory universe* in which observers are *required* to bring the universe into existence… The critical role of life and consciousness in shaping the universe becomes clear."

Lanza concludes, "So you either have an astonishingly improbable coincidence revolving around the indisputable fact that the cosmos could have any properties but happens to have exactly the right ones for life or else you have exactly what must be seen if indeed the cosmos is biocentric [that there is no independent external universe outside of biological life]. Either way the notion of a random billiard-ball cosmos that could have had any forces that boast any range of values, but instead has the weirdly specific ones needed for life, looks impossible enough to seem downright silly." This is a central and critical concept which, upon consideration, must lead us, unarguably, toward an understanding of our existence far different than the standard assumption in our culture that the entirety of existence is physical and material.

Lanza finishes his point, "And if any of this seems too preposterous just consider the alternative, which is what contemporary science asked us to believe: that the entire universe, exquisitely tailored to our existence, popped into existence out of absolute nothingness. Who in their right mind would accept such a thing?" (Lanza, 2009, p. 91).

Numerous astronomers as well have weighed in on this perplexing synchronicity—and with unabashedly mystical appraisals:

For example, Fred Hoyle, a British astrophysicist, credited with naming the big bang concludes: "A common sense interpretation of the facts suggests that a 'superintellect' has monkeyed with physics, as well as with chemistry and biology, and that there are no blind forces worth speaking about in nature. The numbers one calculates from the facts seem to me so overwhelming as to put this conclusion almost beyond question." (Hoyle, 1982, p. 16)

Alan Sandage, who is a winner of the Crawford Prize in astronomy, considered by some to be the greatest and most influential observational astronomer of the last half-century, mused that "I find it quite improbable that such order came out of chaos. There has to be some organizing principle" (Goot and Vanderbei, 2010, p. 89)

Additionally, NASA astronomer John O'Keefe, who is credited with the discovery that the earth is actually pear-shaped, resulting in a larger surface area in the southern hemisphere, takes the perspective that: "If the universe had not been made with the most exacting precision we could never have come into existence. It is my view that these circumstances indicate the universe was created for (humanity) to live in" (Heeren, 1995, p. 200).

Irrefutably, scientists have identified scores of calculations regarding the forces and speeds and charges in the first moments of the universe, along with the chemicals and gases formed out of the earliest atoms, which, had any of it varied in amount, power, or interaction by even miniscule amounts, the universe, let alone life or us, would not have happened! The symphony would have gone flat!

We now know, for example, that:

- If the *speed of the expansion* of the universe was even a bit faster or slower *the universe would not exist:* either it would have collapsed into nothingness on the one hand, or have been stripped of stars and galaxies on the other.
- If *the force of gravity* that shapes and guides and structures the

universe and everything in it were a tiny percentage stronger, *the Earth and the Sun would not exist.*

- Additionally, even a very small adjustment in *the electrical charge of an atom's electron* would have resulted in *an impossibility of chemical reactions occurring.*
- And if the *attractive force of protons in atoms* were a small percentage higher, all hydrogen would have turned into helium, and *no matter would have emerged.*
- Without *the four known forces* in the universe having somehow come into existence, we know with certitude *we would not be here.*
- At the end of the day, each of these precise realities would have had to occur in concert with each other for the universe, for life, and for us to have come to be! (Rolston, 2010, pp. 17-19).
- There are countless other specific tweaks in the physics and chemistry of the universe as well, that, had they been slightly different—each one, let alone the collection!—*this universe and all life would not have come into being.* Underscoring this baffling reality, Walker cites cosmologist Guth's observation: "for the universe to have made it this long, some 14 billon years, requires a fine tuning of the density of matter to 1 part in 10^{33}! This requires knife-edge 'adjustment' of the initial conditions of the universe" (Walker, 2000, p. 323).

Notes Lanza, summarizing, there are "just more than two hundred physical parameters within the solar system and universe so exact that it strains credulity to propose that they are random—even if that is exactly what standard contemporary physics baldly suggests. These fundamental constants of the universe—constants that are not predicted by any theory—all seem to be carefully chosen, often with great precision, to allow for the existence of life and consciousness…." (Lanza, 2009, p. 7).

A further distinguished cohort of experts in various disciplines come to similar conclusions, suggesting that there is much more to the universe than traditional materialistic science has allowed: a driving force, an intelligence, a visionary conception of some sort.

Roger Penrose, well-regarded mathematical physicist, mathematician, and philosopher of science, known particularly for his contributions to general relativity and cosmology, who also explored the connection between fundamental physics and human and animal consciousness, exuberantly proclaimed, "I would say the universe has a purpose. It's not there just somehow by chance" (Morris, *A Brief,* 1991).

Arthur Eddington, an astrophysicist who is best known for his observational confirmation of Einstein's General Theory of Relativity unabashedly ventured the notion that "the idea of a universal mind [emphasis added]…would be, I think, a fairly plausible inference from the present state of scientific theory" (Heeren, 1995, p. 233).

Further, Arno Penzias, an American physicist and Nobel laureate who is co-discoverer of the cosmic microwave background radiation, which helped establish the big bang theory of cosmology reasoned that "astronomy leads us to a unique event, a universe which was created out of nothing, one with the very delicate balance needed to provide exactly the conditions required to permit life, and one which has an underlying (one might say 'supernatural') plan" (Margenau and Varghese, 1992, p. 83).

Theoretical astrophysicist George Greenstein pondered the following conception: "As we survey all the evidence, the thought insistently arises that some supernatural agency—or, rather, Agency—must be involved. Is it possible that suddenly, without intending to, we have stumbled upon scientific proof of the existence of a Supreme Being? Was it God who stepped in and so providentially crafted the cosmos for our benefit?" (Greenstein, 1988, p. 27).

Considered to be one of the world's leading theorists in cosmology, who, with Stephen Hawking, co-authored *The Large Scale Structure of Space-Time*, astrophysicist George F. Ellis marveled, "amazing fine-tuning occurs in the laws that make this [complexity] possible. Realization of the complexity makes it very difficult not to use the word 'miraculous'" (Ellis, 1993. p. 30).

Physicist Tony Rothman, who made significant contributions to the study of cosmic nucleosynthesis, black holes, inflationary cosmology, and gravitons was enticed into wondering, "when confronted with the order and beauty of the universe and the strange coincidences of nature, it's very tempting to take the leap of faith from science into religion" (Casti, 1989, pp. 482-483).

Vera Kistiakowsky, an expert in observational astrophysics and experimental particle physics, reflected that "the exquisite order displayed by our scientific understanding of the physical world calls for the divine" (Margenau and Varghese, 1992, p. 52).

Paul Davies, an astrophysicist, writer, and broadcaster, commented that "there is for me powerful evidence there is something going on behind it all....It seems as though somebody has fine-tuned nature's numbers to make the Universe....The impression of design is overwhelming" (Davies, 1998, p. 203), and "The laws [of physics]...seem to be the product of exceedingly ingenious design... The universe must have a purpose" (Davies, 1984, p. 243).

Idit Zehavi, an astrophysicist and researcher who discovered an anomaly in the mapping of the cosmos, which resulted in insights into how the universe is expanding, along with cosmologist Avishai Dekel, who is known for contributions to research in cosmology, notably his work on the formation of galaxies and large-scale structure in the universe, together wrote: "This type of universe...seems to require a degree of fine tuning of the initial conditions that is in apparent conflict with 'common wisdom'" (Zehavi and Dekel, 1999, pp. 252-254).

Noted for his work on the increase of fluctuations in the expanding universe, astronomer and cosmologist Ed Harrison challenged: "Take your choice: blind chance that requires multitudes of universes or design that requires only one.... Many scientists, when they admit their views, incline toward the teleological or design argument" (Harrison, 1985, p. 263).

Well-known British theoretical physicist, Stephen Hawking, who was the first to propose a theory of cosmology integrating general theory of relativity and quantum mechanics, quite poetically states: "... We shall...take part in the discussion of the question of why it is that we and the universe exist. If we find the answer...it would be the ultimate triumph of human reason—for then we would know the mind of God" (Hawking, 1988, p. 175).

Finally, Max Planck, quantum physicist, Nobel Prize winner, and an originator of quantum theory, confidently asserted that "we must assume, behind this force, *the existence of a conscious and intelligent mind* [emphasis added]. The mind is the matrix of all matter" (Sullivan, January 25, 1931, "Interviews with").

Each of these venerated and accomplished scientists, looking directly at the concrete data from science, concludes that the universe is far more abstract than actual, far more miraculous than material! Should we doubt what they are concluding? Shall we challenge them and remind them that the dogma of traditional science asserts that the only things that exist are physical and material? That what appears as a transcendent and ethereal symphonic masterpiece, a cosmic mind, is merely a random assemblage of nondescript, inert, ordinary, and unremarkably lifeless, mechanical, and insensate atoms?

A MOTHER UNIVERSE?

Any way we conceive of the beginnings of existence leaves us scratching our heads. The likelihood that we could be here, alive, present, and aware, now, is a consideration for which we have no easy answers. "The chance that a livable *particular* universe like ours would be created is far less than the chance of randomly picking a single atom out of all the atoms in the universe," conclude physicists Bruce Rosenblum and Fred Kuttner in their book, *Quantum Enigma: Physics Encounters Consciousness,* (Rosenblum & Kuttner, 2011, p. 264). It turns out that it is not only sages and saints and gurus who attribute something deliberate, purposeful, and intelligent in the formation and existence of the universe. It is also our most esteemed scientists, who have studied its physics and its dynamics, and from that, conclude the same.

Given the staggering coherence, sophistication, apparent purposefulness, and seamless functioning that describes our universe, there may be another, perhaps more illuminating perspective we might consider as to the nature of the beginning of our universe: that our universe emerged out of a mother universe, a meta-universe or a Metaverse.

In his 2007 book, Ervin László writes,

"The phenomenon cosmologists call the 'fine tuning of the universal constants' is particularly vexing. The three dozen or more physical parameters of the universe are so precisely adjusted that together they create the highly improbable conditions under which life can emerge on Earth…

These are all puzzles of coherence, and they raise the possibility that this universe did not arise in the context of random fluctuations of the underlying quantum vacuum. Instead it may have been born in the womb of a prior 'meta-universe.'

A number of physical cosmologies…harbor the promise of overcoming the puzzles posed by the coherence of this universe, including the mind-blowing serendipity that its physical constants are so finely adjusted…This has no credible explanation…the vacuum fluctuations that set the parameters of an emerging universe would have to have been randomly selected: there was 'nothing there' that could have biased the serendipity of this selection…

The coherence of our universe tells us that all its stars and galaxies are connected in some way. And the fine tuning…suggests that at its birth the vacuum in which our universe emerged was not randomly structured. A previous universe is likely to have informed the birth of our universe, much as the genetic code of our parents informed the conception and growth of the embryo that grew into what we are today" (László, 2007, p. 38).

László is arguing that the big bang theory, its implications, and its unanswered problems don't adequately answer the orchestration of our universe's origins nor its startling attunement. He suggests there is a history, an evolving, and in that process, a particular and cumulative set of attributes that coalesce into the particular world we are given. László reasons that in addition to matter and energy, that "at the roots of reality, there is...also a subtler but equally fundamental factor, one that we can best describe as active and effective information: 'in-formation.'[29]

"In-formation ...links all things in the universe, atoms as well as galaxies, organisms the same as minds," László contends (László, 2007, p. 2). He proposes that out of a mother universe, vast numbers of universes have been born, persisted, and died, and that in-formation evolved within that history. All the in-formation from history was transmitted to our universe in its birth, explaining both the moment we call the big bang, and explaining why we see such richness and such fine-tuning in the initial conditions.

"Cosmologies of the Metaverse are in a better position than the Big Bang theory (which is limited to our universe), to speak of conditions that reigned before, and will reign after, the life cycle of our universe. *The quantum vacuum, the enduring in-formation and energy sea that underlies all 'matter,' did not originate with the Bang that produced our universe,* [emphasis added] and it will not vanish when the particles created by that explosion fall back into it," he says (László, 2007, p. 83).

He holds that, "If we assume that at its birth our universe was fully isolated from other universes we cannot find a natural explanation for its astonishing propensity to bring life forth...'

"Instead of marveling at this improbable scenario and giving up on a scientific explanation of it, we can contemplate the possibility that at its birth our universe was in-formed by a universe or universes that existed prior to it...In the cycle of universes in the Metaverse each universe is in-formed by its predecessor and in turn in-forms its successor...

"Thus, at its origins our universe did not come by its fine-tuned properties by chance: it 'inherited' them from a prior universe," (László, 2007, pp. 85-86), just as in biological evolution, species transcend and include capacities from their progenitors.

The Metaverse proposal does not answer the ultimate "first cause" question. László says, "The Metaverse was there before all universes—its

---

29   When László says "in-formation" he is using a play on words implying an action of being formed as well as information as data. Just as quantum theory says that nothing exists without observation, we can say that emerging information doesn't exist without consciousness to observe and use it.

vacuum was primeval, virginal. How then were the initial conditions of the Metaverse created—*by what*…or is the question *by Whom?* This is the greatest and deepest mystery of them all—the mystery of the origins of the universe generating process itself" (László, 2007, p. 86). László's notion of a superseding Meta-universe points to an evolution of the cosmos that somehow "predated" our universe, and at the big bang, ushered in our universe—all of which underscores and reiterates the passionate challenges to the standard model being offered by scientists and others.

We began this chapter by pointing out the illusory nature of reality, that existence is ultimately composed of non-physical fields, all of which is a Consciousness Paradigm point of view. From there, we described how we currently understand our universe as a physical entity, but still as an expansive, inconceivable, and largely ungraspable story, especially when we approach our understanding from the assumptions of the Materialistic Paradigm. In the ensuing chapter, we will explore quantum theory, and see it is the ultimate "bedrock" of reality and existence, and too, the basis of the Consciousness Paradigm.

Chapter Two:

———— ∿ ————

QUANTUM— THE UNDERLYING STRUCTURE OF THE COSMOS

ILLUSORY MATTER: FIELDS OF VIBRATING ENERGY

"In short, the world is made up of processes, events rather than things."

—Christian de Quincey (de Quincey, 2005, p. 52).

"Activity is the very essence of being."

—Fritjof Capra (Capra, 1982, p. 92).

We stumble as we try to quantify, comprehend, or describe in words the majesty that is our inestimable universe, or, conceivably, our metaverse. We do so as well when we ponder the smallest of what exists in this place. What do we know about its tininess? Despite our profound awe at the impenetrable riddle we recognize in the massiveness of the universe we call our home, the micro components hold a truly mind-bending mystery for us as well. Since Einstein determined that matter and energy are the same thing, scientists have been searching deeper and deeper into the minuteness of the universe to try and determine the slightest thing out of which all else is made. What have they have found?

Nothing!

Incredibly, we now know that matter itself is an illusion. Its smallest components are, as Max Planck described, "packets of action," not hard material: "As a man who has devoted his whole life to the most clear-headed science, to the study of matter, I can tell you that as a result of my research from the atoms, this much: *There is no matter as such* [emphasis added]!" (Planck, *Das Wesen,* 1944). Indeed, quantum theory has revealed that every atom is 99.9% empty space. Duane Elgin, globally recognized speaker, social visionary, and author of numerous books, describes it visually: "If a typical atom were expanded to the size of a football field, the seemingly 'solid' portion (the nucleus) would be no larger than a baseball" (Elgin, 1993, p. 278).

He goes on to clarify, however, that: "So-called empty space is no longer viewed as a featureless vacuum…Space is not an absence of form, waiting to be filled out by matter; instead space is *a dynamic presence* [emphasis added] that is filled with incredibly complex architecture… Space is not static emptiness, but a continuous opening process that provides the context for matter to manifest" (Elgin, 1993, p. 278). This "dynamic presence" is the existence and energy of endless subatomic particles in motion, ever manifesting and ever annihilated.

Further, if we examine the content of the atomic nucleus, we find, at its most minute, only quarks: "unimaginably tiny knots of energy." And the electrons that rotate around the nucleus? The protons and neutrons within the nucleus? Not matter, but packets of energy—packets of action—as well. Mostly space with extremely tiny energy "clusters" whizzing about.

Physicist Max Born ratifies this new, counterintuitive picture of what matter is: "We have sought for firm ground and found none. The deeper we penetrate, the more restless becomes the universe; all is rushing about and vibrating in a wild dance" (Born, 1935, p. 277).

Planck, again, describes the reality of matter:

"All matter originates and exists only by virtue of a force which brings the particle of an atom to vibration, and holds this most minute solar system of the atom together" (Planck, *Das Wesen,* 1944).

Lanza describes a "…Seeming emptiness [that] seethes with almost unimaginable energy, which manifests as virtual particles of physical matter, jumping in and out of reality like trained fleas. The seemingly empty matrix upon which the storybook of reality is set is actually a living, animated 'field,' a powerful entity that is anything but empty" (Lanza, 2009, p. 117).

De Quincey, Capra, Elgin, Born, Planck, and Lanza are all describing reality as living, as a unitary energetic environment, a cosmic web that is full—full with virtual particles, but not filled with matter. "Discoveries in the…sciences of quantum physics, as well as in certain areas of psychology, neurology, evolutionary biology, and consciousness studies are ushering in…a new worldview…" concludes de Quincey, in his book *Radical Knowing.* "In this approach, the fundamental 'stuff' of the world is not solid, substantial, material things—like particles, atoms, and molecules—it is something much more like fields of vibrating energy, or vortices in dynamic warps of space-time. In short, the world is made up of *processes,* events rather than things" (de Quincey, 2005, p. 52).

Lanza addresses the question of how something not material can feel so hard, so physical. He explains what is going on at the subatomic level when we experience solidness instead of some process: "What about if you touch something solid? Push on the trunk of a fallen tree and you feel pressure… That sensation of pressure is caused not by any contact with a solid, but the fact that every atom has negatively charged electrons in its outer shells. As we all know, charges of the same type repel each other, so the bark's electrons repel yours, and you feel this *electrical repulsive force* stopping your fingers from penetrating any further. Nothing solid ever meets any other solids when you push on a tree. The atoms in your fingers are each as empty as a vacant football stadium in which a single fly sits on the 50-yard line. If we needed solids to stop us (rather than energy fields), the fingers could easily penetrate the tree as if we were swiping at fog" (Lanza, 2011, p. 22). [30]

Our experience of hardness is an illusion. There is nothing physical

---

30   Duane Elgin, inspired by Bill Keepin, uses another analogy of an airplane propeller to visualize why we feel things are hard or solid? "When a propeller is spinning, it creates a circular arc that appears to be a nearly solid surface. If we try to put something in its path…it functions as if it were solid."

to "hold" the property of hardness. Lanza points out another common illusion we all have experienced: "Consider an even more intuitive example—rainbows. The sudden appearance of those prismatic colors juxtaposed between mountains can take our breath away. But the truth is we are absolutely necessary for the rainbow's existence. When nobody's there, there simply is no rainbow" (Lanza, 2011, p. 22).

Brian Whitworth, of Massey University in Auckland, New Zealand, and an elaborator of the notion of quantum realism—that is, that the universe is, in truth, a virtual reality—affirms plainly, "Quantum reality consists of dynamic events not static things."[31] The static "things" we experience aren't really there!

Although relativity theory and quantum theory are inconsistent with each other, (except in the theoretical mathematical constructs of string and M theory), both, incredibly, agree that it is not "things" or "matter" that make up physical reality. "Since the quantum is a 'packet' of *action*...events or processes are now understood to be the fundamental nature of physical reality," asserts de Quincey (de Quincey, 2002, p. 24). What we "know" to be physical reality turns out to be not physical at all!

Further, Einstein's relativity theory demonstrated unequivocally that both time and space are relative to the observer: neither is absolute, and this data provides us with more evidence of the illusory nature of what we perceive as reality. "Relativity theory...has made the cosmic web come alive," says Fritjof Capra, theoretical physicist, systems theorist, author, and proponent of living systems theory, "by revealing its intrinsically dynamic character, and by showing that activity is the very essence of being... The universe is seen as a dynamic web of inter-related events." (Capra, 1982, p. 92).

'In her book, *The Field,* McTaggart quoted the esteemed physicist,

---

31 Whitworth, Brian, *Quantum Realism*, (2014), an online paper: file://brianwhitworth.com/BW-VRT1.pdf. Brian Whitworth, Ph.D. is an information systems lecturer with a background in math, psychology, and information systems. His research has focused on the interface between technical and social issues, and on human and computer Information processing. In 2007, he published "The Physical World a Virtual Reality," arguing that the world may be a virtual one, as it offers answers to a number of key conundrums of quantum theory, including resolving the big bang question (via the "boot up" process of a virtual reality, as well as addressing the speed of light limitation, information conservation, non-locality, complementarity, uncertainty, and digital equivalence, among others. Later in this book, (Chapter 6), we will talk about virtual reality as a viable explanation of the phenomenon of quantum mechanics in conjunction with some other theoretical propositions.

Hal Puthoff:[32] "Every quantum physicist is aware of the Zero Point Field.[33] What we tend to think of as a sheer void if all of space were emptied of matter and energy and you examined even the space between stars is, in subatomic terms, a hive of activity."

"The Uncertainty Principle, developed by Werner Heisenberg, one of the chief architects of quantum theory, implies that *no particle ever stays completely at rest, but is constantly interacting with all subatomic matter* [emphasis added]. It means that the basic substructure of the universe is a sea of quantum fields," she says (McTaggart, 2008, p. 19).

Who *we are* and the complete environment in which we exist—indeed the whole universe, the whole cosmos—*consists of forces in motion, of movement.* "We live in a world of energy—we are surrounded by it, we consume it, we transform it, and in a way, *we are it,*" says de Quincey. "The entire world around us," muses De Quincey, "—the land, mountains, forests, deserts, oceans, cities, sky, and the vast, perhaps infinite, cosmos of innumerable stars and galaxies—is a world of energy, vortices, fluxes, flows, currents, and vibrations" (de Quincey, 2005, p. 226).

At the most basic level, even living things are quanta—packets of energy—in a constant state of informational flux. "As Einstein showed, all matter is a form of energy ($E = Mc^2$)," McTaggart says. "On our most fundamental level, living beings, including human beings, [are] packets of quantum energy constantly exchanging information with this inexhaustible energy sea. Living things [emit] weak radiation, and this [is] the most crucial aspect of biological processes. Information about all aspects of life, from cellular communication to the vast array of

---

32 Hal Puthoff is the Director of the Institute for Advanced Studies in Austin, presently is involved in research in theoretical studies of gravitation, inertia, cosmology and energy research, laboratory studies of innovative approaches to energy generation and space propulsion, has worked with the National Security Agency, Stanford University, SRI International, NASA, and other research programs, and has written numerous publications concerning remote viewing, lasers and quantum zero-point-energy effects, a textbook on electron-beam devices, and has patents in the laser, communications, and energy fields.

33 McTaggart: The Zero Point Field is "an ocean of microscopic vibrations in the space between things..." (McTaggart, 2008, , p. xxvii). "Also referred to by physicists as 'the vacuum,' the Zero Point Field was called 'zero' because fluctuations in the field are still detectable in temperatures of absolute zero, the lowest possible energy state, where all matter has been removed and nothing is supposedly left to make any motion. Zero-point energy was the energy present in the emptiest state of space at the lowest possible energy, out of which no more energy could be removed...." (McTaggart, 2008, pp. 19-20).

controls of DNA, [is] relayed through an information exchange on a quantum level…These subatomic particles [have] no meaning in isolation, but only in relationship with everything else… You could only understand the universe as a dynamic web of interconnection," asserts McTaggart (McTaggart, 2008, p. xxvii).

We are having to come to terms with "a wild and different set of rules. Even if you try to remove every atom and every particle, empty space is…alive with activity. Particles are continually popping in and out of existence, erupt out of nothingness, quickly annihilate each other and disappear," says physicist Brian Greene. In fact, space "is so flooded with activity, it can force objects to move" (McMaster, *Fabric*, 2011).

Space is alive with activity. It "is filled with fluctuating fields full of all sorts of jittery things going on. It's a place where particles are constantly fluctuating and annihilating each other," notes Greene. "Empty space is not nothing. It's something." Greene also observes, "In fact, space is so real it can bend, can twist, and it can ripple. It's so real, empty space itself helps shape everything around us, and forms the very fabric of the cosmos."

"…All elementary particles interact with each other by exchanging energy through other quantum particles," says McTaggart, "which are believed to appear out of nowhere, combining and annihilating… in less than an instant—10^{-23} seconds, to be exact—causing random fluctuations of energy without any apparent cause. The fleeting particles generated during this brief moment are known as 'virtual particles.' They differ from real particles because they only exist during that exchange—the time of 'uncertainty' allowed by the uncertainty principle…

"This subatomic tango, however brief, when added across the universe, gives rise to enormous energy, more than is contained in all the matter in the world" (McTaggart, 2008, pp.19-20).[34] "What quantum calculations show," she states, "is that we and our universe live and breathe in what amounts to a sea of motion—a quantum sea of light" (McTaggart, 2008, p. 21).

---

34 "Also referred to by physicists as 'the vacuum', the Zero Point Field was called 'zero' because fluctuations in the field are still detectible at absolute zero, the lowest possible energy state, where all matter has been removed and nothing is supposedly left to make any motion.

Walker expresses the manifestation of matter in this way: "Matter is…discrete energy packets that dart from place to place in a frenzy of quantum jumps, that ebb and flow in waves of chance. It is a world where nothing stays for long where it should be but only stays where it could be. The atoms dance and the electrons hop, zipping about from the wildly random thermal commotion and the blurry speed with which probability waves speckle the tiny electrons around their nuclei, painting fleeting mazes into solid things like you and me…." (Walker, 2000, p. 65).

There is a lot going on around us, (and within us), affecting everything we experience, yet it is invisible to our naked eye. Not surprisingly, then, our scientific understanding of the concept "space" can be confusing to us, as it has evolved over time, and has had changing meanings.

László describes how our perception of "space" has developed: "Vacuum in its ordinary usage means empty space," he says. "In cosmology it is used to refer to cosmic space in the absence of matter…

"Nineteenth century physicists speculated that cosmic space is not truly empty: it is filled with the invisible energy field called *luminiferous ether* [emphasis added]…But this was a short-lived belief… [The] Michaelson-Morley experiments failed to observe the expected effect, and the ether was removed from the physicists' world picture. The absolute vacuum—space that is truly empty when not occupied by matter—took its place.

"But the concept of space as empty did not dominate for long… Subsequent observations and experiments showed that *this matrix has a physical reality of its own* [emphasis added]…Thus the vacuum of space is neither empty space nor a pure geometrical structure: it is a physically real medium that interacts with matter and produces physically real effects…It is not just the geometry of space-time, but a real physical field [the zero-point field] producing real physical effects" (László, 2007, pp. 61-62).[35]

---

35 László clarifies the universal field concept (László, 2007, p. 75): "The history of the field concept demonstrates that when phenomena occur that require a physical explanation, scientists first attempt to give an explanation specifically related to the entities that manifest the phenomena. As theories grow and develop, the explanatory concepts tend to become more general. In this way, what were initially seen as local force fields are later understood as universal fields, present at all points in space and time. Electrical and magnetic phenomena are now ascribed to the universal EM-field, the mutual

A Universe Full of Magical Things

László concludes: "The vacuum turns out to be a cosmic medium that transports photon-waves (light) as well as density-pressure waves, exerts the force that may ultimately decide the fate of the universe, and endows the particles we know as 'matter' with mass. Such a medium is not an abstract theoretical entity. It is not a vacuum, but a physically real and active *plenum*" (László, 2007, p. 66).

This quantum sea, the zero-point field, is, unarguably, what underlies our being—farfetched, inconceivable, and incredulous as it is to us all. It is a quantum world we live in, which we translate into symbols—an illusory physical reality.

Two Conflicting Paradigms?

"Scientific materialism is usually confused with science itself. By presenting itself as science, scientific materialism acquires the prestige of science, while avoiding scrutiny into those areas where they differ. The influence of scientific materialism is therefore both pervasive (like science) and subtle to the extent that we are unaware of how much we are under its spell."

—B. Alan Wallace (Wallace, 2008, p. 86).

Simply stated, scientific materialism assumes that existence is in its entirety made up of physical entities. This premise is juxtaposed against an opposite and entirely contrary conception of reality, quantum dynamics, which holds that existence consist of nothing physical! As de Quincey says, "We are thoroughly embedded in this vast energy matrix as it streams in, through, and around" (de Quincey, *Synchronicity*, 2015).

Bruce Lipton and Steve Bhaerman reiterate this, and attest to its present day acceptance in modern science: "Einstein showed that atoms are not actually made out of matter, but consist of non-material energy! Today, it is fully established that physical atoms are comprised of a menagerie of subatomic units, such as quarks, bosons, and fermions.

---

attraction of non-contiguous objects is ascribed to the universal G-field, and the presence of mass is ascribed to the universal Higgs field." And to that we can add the field of fields, the zero-point field.

Interestingly, particle physicists perceive of these fundamental atomic units as vortices of energy resembling nanotornados" (Lipton and Bhaerman, 2009, p. 99).

As unimaginable as this quantum world turns out to be, it is as immensely practical and contributory to our technological landscape as it is unfathomable. "Out of quantum theory has occurred virtually all of the information technology, communication, transportation, entertainment, scientific discoveries, music, video, construction, medicine, and so on and on that we live each day…Quantum mechanics is stunningly successful. Not a single prediction of the theory has ever been wrong. One-third of our economy depends on products based on it," state Rosenblum and Kuttner (Rosenblum and Kuttner, 2011, p. xi). What is real is quantum packets of energy. What we experience is somehow akin to *The Matrix!* (Silver, *The Matrix*, 1999). [36]

Growth across the board in the sciences has been exponential since the early 20th century. This informational explosion has occurred in astrophysics, molecular biology, cosmology, chemistry, botany, neuroscience, nanotechnology, computer science, epigenetics, and now, quantum optics, quantum chemistry, quantum computing, and quantum cryptography, to name only a few. Additionally, because of the exciting and continually expanding potentials of relativity and quantum theory, along with the notion of a holographic universe, complexity and chaos theory,[37] string and superstring theories, M-theory, and so on—the science many of us were raised with, the "old" Newtonian material science of celestial mechanics, has been modified, enhanced, and altered so that old habits of thinking are in many ways unrecognizable.

---

36 According to Wikipedia, "*The Matrix* is a 1999 science fiction action film written and directed by The Wachowski Brothers and starring Keanu Reeves, Laurence Fishburne, Carrie-Anne Moss, Hugo Weaving, and Joe Pantoliano. It depicts a dystopian future in which reality as perceived by most humans is actually a simulated reality called "the Matrix", created by sentient machines to subdue the human population…" Brian Whitworth differentiates the movie's theoretical construct from a quantum construct: "This theory is not that idea, that our physical world is created by another physical world. In virtual realism, the quantum world that creates ours is not physical. Quantum theory tells us that quantum states appear and disappear in a way that physical states cannot, tunnel past barriers no physical particle can pass, ignore speed of light limits when entangled and superpose as physical states cannot, e.g. a quantum current can go both ways round a circuit at once. What quantum theory describes is in every way physically impossible, so physicality cannot be its base (Whitworth, 2014, *Quantum*).

37 Which helps us work with weather, ecology, acoustics, epidemiology, electrochemistry, and fractal geometry applications.

A Universe Full of Magical Things

Walker describes these old habits of thinking, observing that "The power of the current [culturally orthodox, Newtonian] materialist paradigm of physics (and of science generally) is founded on the principle that nothing exists but objective reality, nothing but the physical world, nothing but pieces of matter…It would seem ultimately to explain everything in terms of the materialist philosophy" (Walker, 2000, p. 5). We'll see that, from a larger perspective, this paradigm cannot be true.

Lanza remarked that "Few pause to consider the limits of logic and language as the tools we generally employ in our quest for knowledge. As quantum theory increasingly gains ascendancy in everyday technological applications, as when we create tunneling microscopes and quantum based computers, those actively working to find applications for its marvelous facets often confront its illogical or non-rational nature but ignore it. After all, only the math and technological applications matter to them. They have a job to do; leave *meaning* to the science philosophers. Moreover, one needn't understand something in order to enjoy its benefits…." (Lanza, 2009, p. 136).

The problem with classical mechanics is summed up by de Quincey: "Newtonian mechanics leads to strict causality and determinism, immortalized by Laplace's famous image of the universe as a gigantic clockwork. Within the Newtonian universe, all events that ever happened anywhere at any time were wholly determined by prior mechanical causes. All causality, therefore, was to be explained in terms of an unbroken chain of mechanical causes and effects. The present was wholly determined by the past, and the future likewise would be merely an extension of the same deterministic chain of events. There was no room for a nonmaterial, nonmechanical mind to step in and influence the causal chain. In such a universe, there could be no room for true creativity. Anything 'novel' was merely the rearrangement of already existing pieces of matter. Everything in the world could be reduced to the mechanical workings of the smallest parts. Complete knowledge of the workings of the parts would yield complete certainty about future events. The promise of Newtonian science, therefore, was the possibility of complete causal explanations of wholly deterministic events, giving us complete certainty. And, once science uncovered all the various mechanical laws, it could fulfill Francis Bacon's early promise of giving us control and mastery over nature" (de Quincey, 2005, p. 20). Or so it seemed.

That paradigm is quite a head scratcher, suggests Whitworth: "An objective space and time should just 'be,' but space contracts and time dilates in our world. Objective things should just inherently exist, but electrons are probability of existence smears that spread, tunnel, superpose and entangle in physically impossible ways. Cosmology now says that the entire physical universe just popped up, out of nothing about fourteen billion years ago. This is not how an objective reality should behave!" (Whitworth, 2014, *Quantum*).

Rupert Sheldrake describes what he sees as the "10 core beliefs that most scientists take for granted." These beliefs—articles of faith and unproven assumptions—are now called into question because of what we've uncovered in quantum dynamics. These specious beliefs hold that:

1. "Everything is essentially mechanical...
2. All matter is unconscious...
3. The total amount of matter and energy is always the same, (with the exception of the Big Bang)...
4. The laws of nature are fixed. They are the same today as they were in the beginning, and they will stay the same forever...
5. Nature is purposeless, and evolution has no goal or direction...
6. All biological inheritance is material, carried in the genetic material, DNA, and in other material structures...
7. Minds are inside heads and are nothing but the activities of brains...
8. Memories are stored as material traces in brains and are wiped out at death...
9. Phenomena such as telepathy are illusory...
10. Mechanistic medicine is the only kind it really works...

"Together these beliefs make up the philosophy or ideology of materialism, whose central assumption is that everything is essentially material or physical, even minds. This belief system became dominant within science in the late nineteenth century, and is now taken for granted. Many scientists are unaware that materialism is an assumption:

they simply think of it as science, or the scientific view of reality, or the scientific worldview. They are not actually taught about it, or given a chance to discuss it. They absorb it by a kind of intellectual osmosis," says Sheldrake (Sheldrake, 2012, pp. 7-8).

Materialist science only offers *untested speculations* about the big bang or the birth of the universe, or the definition, origins, and bandwidth of "life:" the credo of science is as conjectural as is the credo of the Catholic Church. Further, an even greater challenge to conventional science has to do with the reality of consciousness, what it is, and how it interplays with what we experience as a physical world: "Where science, all of it, has failed most has been in its inability to cope with the question of mind—the question of the nature of consciousness— the doorway to the quantum mind," challenges Walker. (Walker, 2000, p. 5).

Science itself has, quite ironically, yet without any doubt, concluded that *the data from quantum mechanics points to a non-physical universe,* a Consciousness Paradigm, even while that is blatantly in conflict with the Materialistic Paradigm of science in which the universe, in its totality, is physical and nothing else: if it's not physical, it doesn't exist.

We have two conflicting paradigms. The Materialist Paradigm paradoxically functions entirely well in our everyday world that seems so physical—despite *the inevitable* conclusion of our existence: that "what is real is quantum packets of energy."

According to Lipton and Bhaerman in *Spontaneous Evolution,* "Since the 1700s, three main tenets of the Newtonian philosophy have shaped how scientists approach their study of the universe:

1. Materialism—Physical matter is the only fundamental reality. The Universe can be understood through knowledge of its visible physical parts. Rather than invoking unseen vital forces or spirits, life is derived from self-reactive chemistry that comprises the body. Simply stated, 'All that matters is matter.'
2. Reductionism—No matter how complex something appears, it can always be dissected and understood by studying its individual components. Simply stated: 'To understand something, take it apart and study its pieces.'

3. Determinism—Occurrences in Nature are causally determined, a consequence of the concept that every action produces a reaction. An outcome can be predicted by the linear progression of discrete events. Simply stated: 'We can predict and control the outcome of natural processes'" (Lipton and Bhaerman, 2009, p. 96).

However, "Einstein's mass-energy equation...acknowledges the unification of energy and matter," Lipton explains, "wherein...Einstein showed that atoms are not actually made out of matter, but consist of non-material energy...

"In other words," Lipton reiterates, "the long-held perception of a Newtonian Universe, made exclusively of physical objects, turns out to be an elaborate illusion! In contrast, Einstein... [showed] that the Universe is one indivisible dynamic whole wherein all physical parts and energy fields are entangled and interdependent."

Lipton continues, "Planck's work also questioned the emphasis on reductionism, which focuses on the individual parts rather than the whole. Planck demonstrated that some events cannot be predicted by linear cause-and-effect reactions, but seem to occur spontaneously as part of an interacting energy matrix called *the field*. Planck's insights emphasized that, in order to understand the nature of the Universe, we must abandon reductionism and, instead, turn to holism, wherein everything interacts with everything else" (Lipton and Bhaerman, 2009, p. 99).

The cause and effect reactions that seem so true in the Newtonian universe while perceptually appearing so unarguable, turn out, in the underlying quantum realm, to be spontaneous occurrences in the intermixing energy environment of an all-pervasive field.

"Quantum physics also dispenses with the notion of determinism," Lipton explains, "which is the doctrine that all events, including human choices and decisions are predicated on a specific sequence of causal reactions that adhere to natural law. Simply stated, determinism proposed that with enough data, we can predict the future" (Lipton and Bhaerman, 2009, p. 99). Heisenberg, another of the founders of quantum theory and developer of the Heisenberg Uncertainty Principal, who we introduced above, showed that *predictability at the quantum level is simply not possible.*

Lipton writes, "Not only is Heisenberg's theory a direct affront to determinism, it also suggest(s) that the existence of matter is, itself, an uncertainty."

Lipton continues, "...focusing on the material realm...is a particularly egregious error in the Universe where *we now understand that the invisible field is the agency that governs matter* [emphasis added]. Or, as Einstein stated, with inimitable simplicity, 'The field is the sole governing agent of the particle.' What Einstein meant was that the field was the universe's energy matrix that governs all matter."

Then Lipton observes, pointing toward the core conclusion of this book about the nature of our universe, "Interestingly, *the invisible energy field that shapes matter, as defined by quantum physicists, has the same characteristics as the invisible shaping fields that metaphysicians define as 'spirit'* [emphasis added] " (Lipton and Bhaerman, 2009, pp. 99-101).

Quantum mechanics is the reality of our universe, and of our existence. Scientists have harnessed the information about quantum dynamics to achieve extraordinary and sophisticated contributions to our know-how and technology, even though they are not able to unpack or describe how QM works. What appears to work on the macro level, Newtonian classical mechanics, turns out to be only a special case in the all-encompassing theory of quantum mechanics as it plays out at the micro—atomic and subatomic—level.

Seth Lloyd offers this spicy perspective: "Quantum mechanics is just counterintuitive and we just have to suck it up" (Wolchover, "Have We," 6/30/2014).

De Quincey elucidates further aspects of quantum theory which we'll amplify on in the next chapter: "Quantum theory... [showed] that subatomic physical events are *not causal.* Quantum events are inherently uncertain and unpredictable...Quantum events are *nonlocal*—unhindered by distance. Furthermore, quantum events also involve the choice or consciousness of the observer (i.e., events that are considered 'objective' are in some way influenced by the observer's subjectivity)...

"The central concept in quantum theory is that of the quantum itself, described as a packet of energy or action. Energy or action, then, is quantized—that is, energy comes in discrete 'bundles' of indivisible packets called 'quanta'—with *nothing in between.* In a quantum universe the world is full of 'gaps.' But these are strange 'gaps': they form the quantum void, an infinite sea of quantum potential, of probability waves, which *when observed* collapsed to form the world of actualities…"

De Quincey continues, "Mechanism cannot be the whole story if there are quantum gaps in the world—because mechanism requires continuity and contact between adjoining objects to transmit the energy of their collisions. Furthermore, if quantum events happen only when *observed,* the supposed objectivity of science that goes hand-in-hand with mechanism comes into question. Clearly, if the subjectivity or consciousness of an observer is somehow responsible for 'collapsing' quantum probabilities into an actual event, objectivity is compromised at a fundamental level" (de Quincey, 2005, pp. 23-24).

Quantum theory recasts mechanistic physics as a subset of a grander, all-inclusive perspective in which the material world does not exist as such, where a unitary field binds all of reality, all of the universe, together, and where who and what we are and what we experience depend entirely on what we observe—i.e. *depend entirely on consciousness*—for their existence. Quantum theory recasts an inert universe as thriving, dynamic, energetic, and living events and processes, evolving and emerging—all of which is the subject of our explorations.

Quantum Roots

It may be helpful here to take a step back for a moment and look at the various ways we western humans have thought about the real world and to offer a bit more context. Before we describe some of the mysterious features of quantum theory, a bit of historical narrative can set the stage.[38]

---

38   The contributions of the Eastern world to human knowledge and evolution are vast and partner as equals with the contributions of the Western world. Nevertheless, it was in western civilization where the concepts and experiments and data for the greater part were developed that have led to this peculiar mind-altering paradigm— quantum mechanics—because of which we have now come to understand what truly describes our "real" world. It is therefore the backdrop of the history of western civilization that provides the context for quantum theory.

A Universe Full of Magical Things

Ancient Greece (454-404 BCE) and the Roman Empire (27-476 CE) are considered to be classical cultures, renowned for their emphasis on curiosity, learning, order, science, art, philosophy, and social structure, among other things. For these classical cultures, it was thought that there were two realms: the Heavenly realm above, the world of the ethereal, considered to be perfect; and the Earthly realm, the material world, below, which was not. Each realm had its own laws and rules.[39]

When the Roman Empire collapsed under the barbarian invasions, Europe experienced cultural and economic deterioration and disorder, and entered the broad-ranging Medieval Period, or Middle Ages, (500-1500 CE).

The early Middle Ages (500-1000) are often referred to as the Dark Ages, in contrast to the High Middle Ages, (1000-1300), which, sparked by changes in climate as well as technological and agricultural advances, and the establishment of the feudal political system, led to higher crop yields and social advances. This period in turn ceded to the Late Middle Ages, (1300-1500), a period again marked by decline, due to famine, bubonic plague (i.e. the "Black Death," 1346-53, in which a third of the population of Europe died), and challenges to the religious, political, and economic authority of the Catholic Church, which had dominated most of Europe politically and spiritually throughout this entire period.

The rigidity of and attempts by the Catholic Church to not only control the politics of Europe, its economy, its thoughts, and its leadership, but the very souls of the people, were bound to be challenged at some point. As early as 1231, the church reacted to increasing rebellious and heretical rumbling by establishing a tribunal, called the Inquisition, for the suppression of any defiance by all means, including torture.

The European Renaissance, (1300-1600) which is considered to have begun in Florence, Italy, signaled a rebirth of learning and a harkening back to the sensibilities and values (in art, science, and philosophy) of the ancient Greco-Roman cultures. This rebirth was fueled in part by Gutenberg's invention of the printing press in 1439, and it

---

39 Richard Wolfson, Great Courses lecturer, from the course "Einstein's Relativity and the Quantum Revolution." This view prevailed until the 1500s—the time of Copernicus, Galileo, Kepler, and Newton—when the first stirrings of the scientific revolution arose.

contributed to significant educational advances.

New thought and new technology and broader access to ideas contributed to the looming threat to Catholic authoritarianism, as did, importantly, the Protestant Reformation (1517-1648). Martin Luther contested the Pope's spiritual authority, declaring that the Bible was the source of spiritual guidance, not the Pope. He was initially wanting renewal within the church, not schism from it, but the collective disruptive forces of the times altered forever the influence of the church in European and western culture.

In 1542, the Inquisition was reestablished specifically to combat Protestantism, and it was under this incarnation that Galileo Galilei was tried. Witch trials and witch burnings were another way the church attempted to impose its orthodoxy, commencing about 1580, and continuing till about 1782—a period during which an estimated 40,000 to 60,000 people were executed.

The unfolding of western culture prior to the Modern Period, (beginning in the 1400s), was marked, as we are describing, by intermittent times of immense curiosity, creativity, social and technological advances, and new knowledge. One theme that continually resurfaced, along with this hunger to more deeply comprehend the world and exert more and more control over it, were alternate or competing influences desiring to impose one version or another of orthodox thinking on the populace, as was experienced by the denizens of ancient Greece and Rome and the faithful of the Catholic Church.

For the Greeks, concerns involved the state versus the individual, gender roles and relationships, the role of the gods in human affairs, a philosophical emphasis on reason, and broad considerations on the meaning of life.

The Roman Empire was united by might and geography and so on, but the glue, to an important degree, was that the Latin language was solidified as the *lingua franca* of the land, and continues its sway today through the romance languages—French, Italian, Spanish, Portuguese, influences in English, and others. Noteworthy within the Roman Empire were features such as a value of inclusive citizenship (with welfare), along with newspapers, engineering of aqueducts and a network of roads and public toilets and baths, and general attempts to limit poverty and debt. The concept of "citizen" was formulated and early elements of democracy were valued, along with a limited dictatorship. The degree of deliberate sensibility concerning inclusion and quality of life for the populace at such an early period in history is remarkable. The Roman Empire is also responsible for the spread of Christianity across Europe and beyond.

Christianity, under the guise of the Catholic Church, served as a constant reminder of a life and world beyond the material plane, bolstering a sense of meaning and value for its believers, at whatever corresponding psychological costs. After the demise of the Roman Empire, the Catholic Church served as a unifying force for Europe. Via its churches and cathedrals, the church made significant contributions to architecture, art, and engineering. The reemergence of the idea of the university arose out of a Christian setting. Most of all, the code of Christian morality had a massive influence across the world, and across cultures, and across time. It has, as well, been a major player in regard to the question of women's value and women's rights in a variety of cultures.

In contrast to the flexibility and fluidity in the Greco-Roman mindset that resulted in an openness about knowledge, an encouragement of curiosity, and a humility about the limits of what we know, the Catholic Church up to, and into, the Modern Age, promoted a fear-based, dogmatic, one-right-way orthodoxy that constrained curiosity, knowledge, creativity, and visionary thinking.

Out of this coercive environment in the Early Modern Period, (1400-1800), inevitably came a rupture of the unyielding political and philosophical structures and strictures of Catholicism. In addition to the churning upsets spawned by the Reformation, the rift was spearheaded in the knowledge arena by individuals like Copernicus, Galileo, Kepler, Descartes, Newton, and others, who upended what was thought to have been true at the time. In the social arena, the Age of Enlightenment (1715-1789) sanctioned at least intellectual rebellion against the old order, placed a premium on reason, and nurtured the seeds of life, liberty, and democratic "general will:" notably, the French Revolution contributed to the rise of nationalism, the end of feudalism, and promoted, among other things, democratic freedom, equal treatment, and a voice for all. This was the environment within which the founders of science, and, ultimately, the forebears of quantum theory, were birthed.

Among the first challenges to the church in the sphere of science, was the contention of Copernicus, (1473-1543), that the sun was the center of the universe and not the Earth. At first, this assertion was not controversial within the Catholic Church, but when the increasingly influential Martin Luther and the Protestants loudly objected because this new scientific documentation was thought to contradict a literal interpretation of the bible, the church took on this view as well.

In 1609, based on his astronomical observations, Galileo, (1564-1642), affirmed the Copernican notion of a heliocentric universe, stirring further controversy, as the bible (and for traditional Catholics, the pope), were considered the authorities on faith and morals, not science. The Roman Inquisition in 1615 claimed that Galileo was attempting to reinterpret the bible, and declared that heliocentrism was "absurd and dangerous in philosophy, and formally heretical" (The Roman Inquisition, 1615). The net result for Galileo was that the final years of his life were lived under house arrest.

Johannes Kepler, (1571-1630), now understood to be a key individual in the 17th century scientific revolution, discovered the three laws of planetary motion, including that planets move in an elliptical orbit around the sun. These discoveries later served as a foundation for Newton's theory of universal gravitation, an essential principle of classical mechanics. Though a devout and loyal Christian, Kepler himself was excommunicated from the church in 1612, having ratified the work of Copernicus and Galileo that the universe was heliocentric, not Earth-centered, or geocentric, as maintained by the church. At one point, as well, his mother, Katharina Kepler, was charged with witchcraft, but eventually released.

René Descartes, (1596-1650), philosopher, scientist, and mathematician, is considered to be the "father of western philosophy" and another one of the key figures of the scientific revolution. His original writings are premised on his refusal to accept the authority of other philosophers and even his own senses, laying the foundation for rationalism, which regarded "reason as the chief source and test of knowledge," (Rationalism, *The Encyclopedia Britannica*) employing deductive reasoning, ("I think, therefore I am"), and rejecting empiricism, that knowledge is reliably derived from sensory experience. He asserted that the body works like a machine, a central insight in Newtonian mechanics, whereas the immaterial mind, non-physical and non-spatial, is a distinct and separate substance. Called "Cartesian Dualism," Descartes thought that though separate, the two substances causally interact. He categorized the mind as including consciousness and self-awareness, as distinct from the physical brain, which he saw as the seat of intelligence.

In 1629 Descartes completed a treatise, *The World,* (Descartes, 1629, *The World*), within which he tried to show that mechanistic physics could explain phenomena without appealing to verification by academic principles, but, because of Galileo's condemnation by the Inquisition, he did not publish the work. His thinking in this area, particularly, served as an influence on Newton.

Over the course of his life, Descartes, emphasizing logic and reason over sensory experiences, sparked the early Enlightenment era with significant contributions to physics, mathematics, optics, philosophy, and even made contributions to psychology and physiology. His philosophical contributions were developed with an intention to extend mathematical methods to all fields of human knowledge. He also had a broad influence on the law and theology of the time.

He distinguished a scientific view ("Of what can we be certain?") from a religious view ("What is true?"), amending the source of authority from what God ordains, to, instead, what humanity can document through reason—providing the Enlightenment and the scientific revolution as well with the final liberation from the orthodoxy of the church, and the death knell to the church's overarching domination over the masses of Europe. The scientific method for determining what is real wrested supremacy from the church's dogmatic credo once and for all, and the early architects of the scientific revolution—Copernicus, Galileo, Kepler, and others were vindicated, transforming them from heretics to heroes of the new Materialistic Paradigm.

Newton, (1642-1727), English scientist, mathematician, and philosopher played an essential and influential part in the scientific revolution. He developed the principles of classical mechanics, referred to also as celestial or "Newtonian mechanics," and presently considered as representing "materialist," "determinist," and "reductionist" science, due to the core principle that all that exists on the Earthly plane is what is physical, as we have discussed above. Newton argued for the superiority of classical mechanics over the system developed by Descartes (with matter and mind coexisting but separate). He made significant contributions in mathematics, and especially in calculus. Taken together, his work resulted in an enormous acceleration of knowledge in the fields of mathematics, physics, astronomy, and optics, and he is considered to have left a legacy comparable only to Einstein.

He developed the three laws of motion, and his own mathematical understanding of gravity. With these discoveries, he effectively established the cause and effect nature of all things physical, demonstrating the mathematical formulae behind the motion of matter. Every billiard ball trickster illustrates the predictability of matter and motion discovered by Newton, and every application of science from Newton's time until the discovery of quantum mechanics—and beyond—utilizes the three laws of motion and the understanding of gravity developed by Newton.

For three hundred and fifty years, science proceeded on the premise of a mechanistic, reductionistic, and deterministic universe, a universe that more and more over time elevated physics and matter as a superior paradigm for understanding the world at the expense of both the inner morality-based world of religion and the nonmaterial mind, i.e. consciousness—the other half of Cartesian dualism. Consciousness, subjective experience, and self-reflective awareness, say the materialist-based scientists and philosophers, are only "epiphenomena" of the brain, something that emerges out of the primacy of a universe that is matter-based, an "emergent property" emanating entirely from the world of material.

For three hundred and fifty years, Newton's perspective was sufficient, and—for all practical purposes—it *is* sufficient to explain and describe and concretely apply the way things work with larger than molecular objects, macro-objects of rocks and trees and worms and trucks and people.

Despite this, as Sheldrake noted, "The belief in determinism, strongly held by many nineteenth and early twentieth-century scientists, turned out to be a delusion. The freeing of scientists from this dogma led to a new appreciation of the indeterminism of nature in general, and of evolution in particular" (Sheldrake, 2012, p. 19).

The curiosity of a collection of undaunted physicists inevitably led to an examination of the smaller and smaller objects in the universe that, *when assembled together,* result in the macro-objects of our world. In that micro-world, they found molecules and atoms and fields and forces and energy, and tiny things inside the atom—protons, electrons, neutrons, leptons, muons, quarks, and bosons—and more!

In 1801, well before quantum theory was developed, Thomas Young, a physician who contributed to the scientific advances of the time in the fields of light, vision, energy, physiology, solid mechanics, as well as language and music, demonstrated through his paradigm altering double slit experiment that the wave theory of light was correct. This experimental design eventually led to the modern double-slit experiment which illustrates that light and matter can take the form of both waves and particles and demonstrates the probabilistic (but not predictive) nature of quantum mechanical phenomena, and more, that measurement (observation) was requisite to how any potential event occurred. The double slit experiment is thought to be an emblematic illustration of the contrast between how things behave at the micro verses the macro levels of the material world.

The earliest direct pioneer of quantum theory is considered to be Michael Faraday who discovered black box radiation in 1838. Following further discoveries by a variety of other physicists in the 19th century,[40] in 1900, the originator of quantum theory, Planck, first theorized that electromagnetic energy could only be transmitted in quantized form (intermittent "packets of action" transmitted in a specific frequency). A few short years later, in 1905, Einstein proved that matter and energy are only different forms of the same thing, and that light, as well, travelled in discrete bundles of energy, or action, called photons.

Neils Bohr (who conceptualized a stable model of the atom that

---

40   Including Kirchhoff, Boltzmann, Maxwell, Hertz, Rydberg, Röntgen, Becquerel, Zeeman, Curie, Soddy, Thomson, Rutherford, and Borgman, among others.

included quantum jumps, and who articulated the Copenhagen interpretation in 1922—more later), Louis de Broglie (who identified the wave nature of electrons in 1924), Schrödinger (who developed from de Broglie's idea, the wave equation of matter—Schrödinger's Equation, in 1925), Born (who described electron waves as "probability waves," in approximately 1926), Heisenberg (who developed the Uncertainty Principle in 1927—that we cannot know both the location and velocity of an atomic particle), and Paul Dirac (who with Schrödinger developed a productive model of atomic theory in 1934), are all considered co-founders of quantum theory, in that they advanced Planck's theory and developed its mathematical equations and applications.

By the 1930s, quantum physics had been applied to the electrical properties of solids, the basic properties of the atomic nucleus, and nuclear fission. In 1964, experiments utilizing very high energy between fundamental particles like protons and electrons to uncover the internal structure of protons and neutrons led to the discovery of quarks, important in part because the experiments applied principles of both quantum physics and relativity.

In the latter half of the twentieth century, quantum physics explained superconductivity, which allowed electrical currents to span a solid with almost no resistance, making the development of computers possible. The development of semiconductors and the computer chip using quantum dynamics has further contributed to the information revolution, regarded to be as significant as the industrial revolution. The exploration continues as well in chemistry, with the creation of a great number of original molecules via quantum bonding, (Rae, 2005), and biology. The emergence of nanotechnology (the quantum manipulation of atoms and molecules) in the 1980's has led to molecular nanotechnology and a variety of sophisticated commercial and industrial applications in medicine, biotechnology, energy and green nanotechnology, military and architectural applications, and many others.

Chapter Three:

―――――∿∿∿――――――

QUANTUM—
THE STRANGE FEATURES OF
QUANTUM DYNAMICS

THE WEIRDNESS OF QUANTUM

*"The quantum paradox, that quantum unreality causes
physical reality, is embodied in the question: 'Can something
that affects real events ... itself be unreal?'"*

—H. Dieter Zeh (Zeh, 2004, p. 4).[41]

Quantum theory maintains that subatomic entities can be revealed
as both a particle and as a wave. One can mathematically compute the
probability of either behavior. *But until we observe the result, we won't
know: observation "creates" the result.* Until the "probability calculation"
resolves into a specific observed outcome—a "state vector collapse" or
"decoherence" or "wave function"—it is referred to as being in "superposition," the collection of all possible results and their probabilities.

---

41   Zeh discovered the phenomenon of decoherence, and one of the developers of the many minds
       interpretation of quantum mechanics.

This observation or measurement, which, it is thought, creates a *particular* outcome at the atomic and subatomic level, "turns out," notes A. I. M. Rae, "to be the most difficult and controversial problem in the interpretation of quantum physics... *We are led to the conclusion that the act of detection is an essential part of the measuring process* [emphasis added]" (Rae, 2005, pp. 186-187).

Lanza says, emphasizing this, that "When studying subatomic particles, the observer appears to alter and determine what is perceived. The presence and methodology of the experimenter is hopelessly entangled with whatever [she or] he is attempting to observe and what results [she or] he gets. An electron turns out to be both a particle and wave, but *how* and, more importantly, *where* such a particle will be located remains dependent upon the very act of observation" (Lanza, 2009, p. 49).

Additionally, a surprising number of other aspects of quantum dynamics are perplexing and are inconsistent with Newtonian mechanics, as we will see, including Heisenberg's uncertainty principle, the phenomena of entanglement, non-locality, complementarity, tunneling, and wave-particle duality, among others. We'll look at what the research has determined and at the resulting implications in the next pages.

The preeminent contemporary understanding, or interpretation, that we have of how quantum dynamics might work—one of several handfuls of possibilities, the one that has the greatest acceptance, and the one which represents physicists' best guess—is called the Copenhagen interpretation, which itself is mysterious and begs further questions. According to the Copenhagen interpretation, for example, quantum dynamics illustrates that there is truly nothing material, as we have noted: packets of action, quanta, are what we find at the lowest levels of the micro-world. And further, things are not predictable or deterministic in the quantum world: they are only probabilities.

In classical mechanics, all events are "caused."[42] Everything exists

---

42   Brian Whitworth in *Quantum Realism:* "Last century the standard model was a simple theory of mass, charge and spin but today it needs isospin, hypercharge, color, chirality, flavor and other esoteric features to work. This theory of sixty two core particles, five invisible fields, sixteen charges and fourteen bosons...has so many ad-hoc properties that if it were a machine, one would have to hand-set over two dozen knobs just right for it to light up. If the standard model is preferred today, it isn't because of its simplicity.

in a cause and effect way, as when one hits a pool ball, it causes the other balls on the table to be affected. In quantum dynamics, events occur, but they are "uncaused," or "self-caused." Newton's predictability model is called into question by the quantum theory conclusion that *nothing* is predictable. The best we can do is determine probabilities for what we might observe.

Quantum theory, according to the orthodox Copenhagen interpretation, says that what we experience in the world is affected by the observer. Mind affects matter. Without an observer, nothing happens. Additionally, there is no objective reality: only our unique individual subjective observation is real.

According to classical mechanics and the world we experience every day, none of this is sensible, and quantum realities are not able to be understood as valid or acceptable from a Newtonian point of view. "The firm reductionist worldview is that, at bottom, there is nothing but whatever is 'down there' at the base of physics plus Einstein's spacetime," says Kauffman (Kauffman, 2008, p. 16).

By contrast, "In…quantum entanglement or nonlocality, there is the strong suggestion that the mind itself is intimately involved in what were once considered purely physical processes," explains Wallace. "Indeed, there is very strong scientific evidence for these 'impossible' situations that contradict common sense. The basic theories of quantum physics do not speculate beyond the data. However, if one is biased, even unconsciously, by the closure principle (that all causes are physical) and other beliefs of scientific materialism, one will not seriously attempt an explanation of quantum relationships—essentially because one doesn't believe such things are real" (Wallace, 2008, p. 90).

From the Newtonian perspective, which we utilize and relate to everyday, the rules in the minute realm of quantum dynamics seem to be incompatible, counterintuitive, and paradoxical. Nevertheless, of course, as we have said, the theory of quantum dynamics is the practical basis for our technological world today and underlies all that we

---

"For this complexity one might expect completeness, but the standard model can't explain gravity, proton stability, anti-matter, quark charge, neutrino mass, neutrino spin, family generations, quantum randomness or inflation. In addition it doesn't explain dark energy or dark matter, i.e. most of the universe... and to explain neutrino mass needs another 7-8 arbitrary constants.

"The standard model of physics is a descriptive model...but nothing physical can explain quantum states" (Whitworth, 2014, *Quantum*).

perceive as physical.

With quantum physics we are compelled to accept twelve game-changing realities that defy our comprehension from a Materialistic Paradigm perspective:

1. *Matter is not solid. It is activity—processes and events—which is the essence of being.*

"One startling discovery made by quantum physicists was that if you break matter into smaller and smaller pieces you eventually reach a point where those pieces—electrons, protons, and so on—no longer possess the traits of objects…Although an electron can sometimes behave as if it were a compact little particle, physicists have found that *it literally possesses no dimension,*"[43] according to Rosenblum and Kuttner (Rosenblum and Kuttner, 2011).

What we perceive as physical objects are, at their most infinitesimal, empty space filled only with darting packets of action, quanta, as we have described. What lies in the gap between these quanta is a quantum universe, a sea of probability waves, that when observed, "collapses" into reality, a reality based in activity, not physicality, as we have seen. Indeed, as Capra has described, "activity is the essence of being" (Capra, 1982, p. 92). (So the illusion of boundaries between objects that are not truly objects at all is false: There are no boundaries because there are, fundamentally, no true objects. *The universe is completely interconnected*—about which we'll say more in #11 below).

2. *We do not live in a world that always functions in a cause and effect way. The world consists of more than matter acting on matter.*

We do not live in a completely cause and effect world, where, when one pool ball hits another at a certain speed, it's predictable where the ball will end up. That's classical mechanics, which seems to work well

---

43 "In principle…any two objects that have ever interacted are forever entangled. The behavior of one instantaneously influences the other, and the behavior of everything entangled with it. Since truly macroscopic objects are almost impossible to isolate, they rapidly become entangled with everything else in their environment. The effect of such complex entanglement generally becomes undetectable. Nevertheless, there is, in principle, a universal connectedness whose meaning we have yet to understand" (Rosenblum and Kuttner, 2011, p.189).

A Universe Full of Magical Things

with large objects. The tiniest aspect of those large objects is not governed by classical mechanics, but by quantum mechanics—where it's not possible to know *both* the speed and the location of an atomic particle.[44] Of course, macroscopic objects such as we are, or our measuring instruments are, are governed by the rules of quantum mechanics just as microscopic "objects" are. For example, a chair is made up of packets of action, as are microscopic particles, not material hardness. But the behavior of entities at the microscopic level doesn't, *for all practical purposes*, (FAPP, as the physicist John Bell says), impact the macroscopic level.[45] Newtonian physics works fine, FAPP, at the macroscopic level, say Rosenblum and Kuttner (Rosenblum & Kuttner, 2011, p. 175).

3. *There is always uncertainty, and the world is indeterminate. We live in a world of probabilities, not predictability.*

In the quantum mechanical world, *there is always uncertainty, unpredictability, and the world is indeterminate.*[46] Quantum particles—elementary particles of energy and light—sometimes act like particles and sometimes like waves (the "wave-particle duality"). Because of this complementary dynamic, physicists talk about quantum particles, (quanta), as "wavicles."[47] This was an essential quantum insight: we

---

44   Heisenberg's Uncertainty Principle.
45   Bell "firmly emphasized the depth of the unsolved quantum enigma. In big, bold letters on the blackboard he introduced his famous abbreviation, FAPP, 'for all practical purposes,' and warned against falling into the FAPPTRAP: accepting a mere FAPP solution for the enigma." *Quantum Enigma*, (Rosenblum & Kuttner, 2011, p. 175). The authors use the expression "quantum enigma" to refer to interpretations that view the observer as creator of what we experience.
46   These realities of quantum science are so far from our normal classical experience that they remain puzzling to scientists. Over the years there have been various ideas proposed to make sense of the findings. The most widely accepted and consensus perspective, as we have noted, is the "Copenhagen interpretation," but there are others, including the "DeBroglie/Bohm interpretation," (involving pilot waves), "the Many Worlds Interpretation," "Spontaneous Collapse," "QBism, "Relational Quantum Mechanics," and "Quantum Realism," among others. All have merits, all have problems, and all are being pursued as a way to arrive at more clarity about a subatomic quantum system that is based in probabilities, observation, and subjectivity, rather than in an objective, deterministic one described in Newton's classical mechanics. We will discuss several of these possible interpretations of the quantum data in the next chapter.
47   Despite this indeterminacy, "Perhaps most astonishing of all is that there is compelling evidence *that the only time quanta ever manifest as particles is when we are looking at them.* For instance, when an electron isn't being looked at, experimental findings suggest that it is always a wave…. Physicist Nick Herbert…says this has sometimes caused him to imagine that behind his back, the world is always 'a radically ambiguous and ceaselessly flowing quantum soup.' But whenever he turns around and tries to see the soup, his glance instantly freezes it and turns it back into ordinary reality. He believes this

only know whether an elementary entity will behave as a wave or as a particle *when we observe it*—thus, the significance of consciousness! László suggests this metaphor: "It is as if the observer, or the measuring instrument, fishes the quanta out of a sea of possibilities. When a quantum is pulled out of that sea, it becomes a real rather than a mere virtual beast—but we can never know in advance just which of the possible real beasts that it *could* become it actually *will* become" (László, 2007, p. 26).

Despite the classical reality we experience wherein things appear predictable and certain to us, it remains the probability rules of the quantum system that nevertheless, undeniably and indisputably, *supersede the rules of the classical systems* of our daily life. It is immateriality in the form of events, processes, and action that come to exist as we observe them, and not a preexisting and hard physicality, that is of what reality consists!

4. *Information and action can occur instantaneously across immense distances.*

Somehow, particles can affect each other instantaneously over unlimited distances through non-physical and non-energy means: (They become entangled, are non-local, and are complementary). "...When two quantum objects...are separated in space so that no signal could possibly pass between them... By changing a property on one...the other was observed to undergo a complementary change—*as if it somehow knew what had happened to the first (object)*. This is called non-locality," notes de Quincey (de Quincey, 2002, p. 28).[48]

Two particles can become "entangled," and somehow communicate with each other instantaneously, in a way that confounds us (because the limit on speed is the speed of light, as Einstein established

---

makes us all a little bit like Midas, the legendary king who never knew the feel of silk or the caress of a human hand because everything he touched tuned to gold. 'Likewise, humans can never experience the true texture of quantum reality,' says Herbert, 'because everything we touch turns to matter.'" (Talbot, 1991, p. 34).

48 Non-locality refers to the seeming capability of objects to instantaneously know about each other's state, even when separated by vast distances. Quantum dynamics maintains that instantaneous action and information transfer (transcending the speed of light), is an aspect of atomic and subatomic reality, in contradiction of classical mechanics.

with his theory of relativity, an absolute limit).[49] Yet somehow apparent communication—and "particle event complementarity"—occurs.[50]

Comments László: "Quanta are highly sociable: once they share the same identical state, they remain linked no matter how far they travel from each other"[51] (László, 2007, p.26). And de Quincey further explains: "Quantum events...defy the presumed impossibility of action-at-a-distance. That is, either quantum events *can* exceed the speed of light (which, per relativity, is impossible), or quantum events are *nonlocal*. That is, they are correlated in such a way as to behave as though they were a unity—even when apparently separated by super-luminal distances.

"In other words, quantum events are connected at some deep level of reality. In [physicist] David Bohm's terminology: At the implicate quantum level, reality is an undivided whole. Everything is connected to everything else [like a virtual computer game]. Quantum nonlocality has been repeatedly confirmed experimentally" (de Quincey, 2002, p. 28).

5. Quantum events are uncaused, that is, self-caused.

Quantum events are uncaused—which logically means, according to some philosophers, that an uncaused event is equivalent to a self-caused event, so that *an uncaused quantum event is the same as a self-caused quantum event.* An observer would not be able to distinguish any difference between a cause-and-effect classical mechanical action, on the one hand, from a purely randomly caused, "uncaused," or "self-caused" event on the other. The term "self-caused" can be understood as saying that there is no cause outside of the quantum event itself that explains its action.

---

49   Entanglement means that if two particles interact, the state of the one particle depends on the state of the other, and the two become "entangled." We don't know the state until we measure or observe one of them. Before we measure they are indeterminate. The two particles are not two particles at all, we could say. They are now a complex system consisting of two particles in an unknown state, which must behave consistently with each other, regardless of the distance between them.

50   Alain Aspect and colleagues were able to use newly developed technology to prove that the connection between two photons is non-local. "Furthermore, as Paul Davis of the University of Newcastle upon Tyne, England, observes, since *all* particles are interacting and separating, 'the non-local aspects of quantum systems is therefore a general property of nature'" (Talbot, 1991, p. 53).

51   "Complementarity" is a concept formulated by Neils Bohr, a founder of quantum theory. Despite the fact that a picture of a particle and a picture of a wave are different, quantum theory needs both: they are complementary to each other.

A quantum, says László, "appears to choose its real states *on its own* from among the virtual states available to it." (László, 2007, p. 26) Similarly, and curiously, we know of no cause outside of the universe itself that brings it into being, maintains it, or explains its actions: that might suggest the universe, or perhaps the Metaverse, is self-caused).[52]

"Quantum events are *not* causal," de Quincey explains. "They are inherently unpredictable. The exact instant when an electron jumps orbit, or when a radioactive particle is emitted from an atom, is entirely random, entirely uncaused. To say that an event is 'entirely uncaused' amounts, logically, to saying it is 'self-caused'; and this, as philosopher Arthur Young argued, is tantamount to saying that the event *chooses*, that is, exhibits consciousness"[53] (de Quincey, 2005, p. 98).

This logical and observational identity means that quantum events have no causal antecedent, and thus, choice is a critical implication of quantum physics.

Says de Quincey making an important clarification, "If some entity, say an electron or a photon, is exercising true self-action, free will, or choice in how it will move, its behavior will be undetermined (by any prior causes) and will be unpredictable...The quantum event would be *self-caused*, not uncaused. It is uncaused only to the extent that no causes external to itself influence its behavior. [No mechanistic physics are involved]" (de Quincey, 2005, p. 25).

6. *Each quantum observation occurs from a unique, personal, and subjective vantage point.*

Unlike classical physics, where the world appears predictable in a cause and effect kind of way— and that we can agree on, and then call it "objective," in quantum physics, what occurs requires a measurement, that is, an observer, with a subjective experience unlike another

---

52 This, keep in mind, is according to the preferred, although not by any means uncontested, Copenhagen interpretation. Particles come into and go out of existence, a point of view in which it is understood that there exists nothing until it is observed. Until then, only mathematical probabilities can be determined.

53 "The one cause that science refuses to admit in any guise is 'final,' which refers to a teleological, goal-directed 'pull' from some end purpose (implying that nature's events unfold with *purpose* or inherent intelligence). These have been categorically dismissed by science as unscientific or unreal" (de Quincey, 2005, p .98).

observer. In quantum physics, the world is not predictable and cannot be determined in advance of observation, and each observation occurs from a unique, personal, subjective vantage point.

Remarkably, we find that in quantum theory because of its requirement that an event be observed before we can confirm its existence, there is no objectivity! Every event is witnessed uniquely by an observer, subjectively. "One of the most challenging—and to scientists committed to objectivity, most disturbing—revelations from quantum science is that no…separation between subject and object exists," observes de Quincey. "The observer is not—and cannot—be separated from the object being investigated. The so-called observer is actually a *participator,* an integral part of the quantum system…It is as if the only way the realm of quantum possibilities can become actual, manifest reality is for an observer to *choose* one of the quantum probabilities; this probability then becomes an actual physical entity, such as an electron or a photon. In short, by *participating* in the quantum system, the 'observer' helps create reality…In some as yet unexplained way, a quantum observer—and that means an observer's consciousness— participates in the creation of physical reality [like a player in a virtual game?]…Manifest reality reflects the intentions and choices of sentient participators. *The entire physical world arises out of collapsed quantum events, and, therefore, is partly created by some participating [subjective] consciousness* [emphasis added]" (de Quincey, 2002, p. 27).

Manifestation requires observation. "Eugene Wigner," who, according to Lanza, was "one of the twentieth century's greatest physicists, stated that it is 'not possible to formulate the laws of consciousness [of the observer].' So when quantum theory implies that consciousness must exist, it tacitly shows that the content of the mind is the ultimate reality, and that only an active observation can confer shape and form to reality—from a dandelion in a meadow to sun, wind, and rain" (Lanza, 2009, p. 81).

7. We cannot predict events in advance of our observation.

Without an observation, it seems, no quantum event occurs. "The most revealing characteristic of quantum physics," says Wallace, "is the role of observer in measurement: it is the act of observation, intimately wrapped up in the view of the scientist—his or her beliefs—that determines the outcomes such as wave or particle and other physical states. It seems that at the subatomic level, the level that supposedly underlies all physical reality, the mind acts as a potent, cooperative force in the creation of reality as we know it. Subatomic particles, the instruments that detect them, laws concerning their existence and expression, mathematics,and the mind all exist in dependence upon one another" (Wallace. 2008, p. 115).

Lanza posits that "Reality—according to the most stringent interpretation of the scientific data—is created by or at least correlative with the observer... Even back in the eighteenth century, Immanuel Kant, ahead of his time, said that 'we must rid ourselves of the notion that space and time are actual qualities in things in themselves... All bodies, together with the space in which they are, must be considered nothing but mere representations in us, and exist nowhere but in our thoughts'" (Lanza, 2009, pp. 116-117).

Lanza further contends that time itself is a product of our observation. "Time's nature is... A biological creation that is solely a practical operating aid in the mental circuitry of some living organisms, to help with specific functioning activities," (Lanza, 2009, p. 99), that "...Time does not exist in the universe independent of life that notices it" (Lanza, 2009, p.109).[54] Continuing, he holds, "Space and time are neither physically nor fundamentally real. They are conceptual, which

---

54   Robert Lanza, regarding the nature of time, asserts: "That time is a fixed arrow is a human construction. That we live on the edge of all time is a fantasy. That there is an irreversible, on-flowing continuum of events linked to galaxies and suns and the Earth is an even greater fantasy. Space and time are forms of animal understanding—period... So there simply is no absolute self-existing matrix out there in which physical events occur independent of life.

"The forward motion of time—of which the movement of the [unidirectional] arrow [of time] is an embodiment—is not a feature of the external world but a projection of something within us, *as we tie together things we are observing*. By this reasoning, time is not an absolute reality but a feature of our minds," explains Lanza.

"In truth, the reality of time has long been questioned by an odd alliance of philosophers and physicists. The former argue that the past exists only as ideas in the mind, which themselves are solely neuro-electrical events occurring strictly in the present moment.

"Philosophers maintain that the future is similarly nothing more than a mental construct, in anticipation, a grouping of thoughts. Because thinking itself occurs strictly in the 'now'—where is time? Does time exist on its own, apart from human concepts that are no more than conveniences for our formulas were for the description of motion and events? In this way, simple logic alone cast doubt on whether there exists anything outside of an 'eternal now' that includes the human mind's tendency to think and daydream" (Lanza, 2009, p. 96).

means that space and time are of a uniquely subjective nature. They are *modes of interpretation and understanding.* They are part of the mental logic of the animal organism, the software that molds sensations into multidimensional objects" (Lanza, 2009, p. 113).

In the Copenhagen interpretation of the quantum world,[55] *a particle cannot be said to exist at all,* except as a mathematical probability, *until it is observed.* If something is not observable, we should not assume it has a reality: "Most of us tend to assume that a particle always has to be 'somewhere' even when it is not being observed, but this is not true in a quantum context: if a particle is in a state where its position is unknown, then to think about it even having a position is... meaningless...It is wrong to think that an electron in an atom is at any single point within it," stresses Rae (Rae, 2005, pp. 182-183). The implication: *if we do not observe it, it does not exist, and so cannot be predicted.* Biocentrism contends that "It's *all* only in the mind... the universe exists nowhere else [emphases added]" (Lanza, 2009, p. 167). So that, when observed, a particle becomes real to us, but only as an aspect of our mind.

8. *What happens in the world is affected by our observation.*

Until there is an observation, potential event prospects remain in a kind of suspended animation, or "superposition." In some way, *what happens in the world is affected by, even created by, a participant observer:* "The observer is not and cannot be separated from the object being investigated...The so-called observer is actually a *participator,* an integral

---

"Physicists, for their part, find that all working models for reality—from Newton's laws and Einstein's field equations through quantum mechanics—have no need for time" (Lanza, 2009, p. 96).

Further, Lanza suggests this notion: "... Instead of time having an absolute reality, imagine instead that existence is like a sound recording. Listening to an old photograph doesn't alter the record itself, and depending on where the needle is placed, you hear a certain piece of music. This is what we call the present. The music, before and after the song now being heard, is what we call the past and the future. Imagine, in like manner, every moment and day enduring in nature always [à la Bohm's Implicate Order].The record does not go away. All nows (all the songs on the vinyl record) exist simultaneously, although we can only experience the world (or the record) piece by piece. We do not experience time in which 'Stardust' often plays, because we experience time linearly" (Lanza, 2009, p. 105).

55   "All the entangled experiments of the past decades point increasingly toward confirming Copenhagen more than anything else" (Lanza, 2009, p. 58).

part of the quantum system" notes de Quincey (de Quincey, 2002, p. 27).[56]

From this, we must deduce that we humans are observer-participators in the unfolding of the universe. Walker states that the data quantum physics arrives at "leads us to the incredible conclusion that *mind, or consciousness, affects matter*" (Walker, 2000, p. 95).

"An observation without consciousness would not be observation…In short: *consciousness creates reality*…Quantum theory compels us to accept that the scientist, as observer, is a *necessary part* of every quantum experiment. The quantum physicist is, therefore, *participatory observer*" (de Quincey, 2005, pp. 164-165).

The observer is not separate from the object being investigated. *The objectivity that defines classical mechanics, in the world of quantum, is replaced by subjectivity.* "How could sentience and subjectivity *ever* emerge from wholly insentient and objective matter?" asks de Quincey.

"… If you begin with purely objective matter, *without the slightest trace of mind or subjectivity,* no amount or degree of complexity will yield mind or consciousness. It is inconceivable that objectivity could ever give rise to subjectivity or 'interiority,' that 'dead' matter could ever produce matter or minds that *feel.* If the universe is dead to start with, it stays dead" (de Quincey, 2002, p. 42).

What I as an observer see, is subjective—what I see in contrast to what you see. We are both the subject. We each take a different vantage point. In fact, because time and space are relative, as Einstein's relativity theory tells us, our observations *MUST* be subjective. De Quincey comments: "If, as both quantum and relativity theory tell us, events, processes, or durations lie at the heart of reality, then reality cannot be wholly objective, cannot be wholly mechanistic—cannot be made up of 'dead' matter…The notion of time, of duration, of process is intimately related to subjective experience. Subjectivity and sentience, not mere mechanism, appear to be fundamental…" (de Quincey, 2002, p. 24).

So not only quantum theory, but relativity theory as well speaks of time and space as changing depending on an observer's position within them. For example, Hawking and Mlodinow declare from the

---

56   John Archibald Wheeler, an associate of Einstein and Bohr, who died in 2008, famously coined the terms "Participatory Anthropic Principle" and "Participatory Anthropic Universe," notions that will inform the conversation ahead.

perspective of Einstein's theory of relativity: "In the theory of relativity, there is no absolute time; instead, each individual has his own personal measure of time that depends on where [each individual] is and how [they] are moving...Space and time not only affect, but are affected by everything that happens in the universe" (Hawking and Mlodinow, 2005, p. 48).

Further, we each have our own "interiority," we each have our own unique consciousness. Donald Hoffman, a cognitive scientist at the University of California (Irvine) notes in an April 21, 2016 interview with *Quanta* magazine, echoing Lanza, that: "The idea that what we're doing is measuring publicly accessible objects, the idea that objectivity results from the fact that you and I can measure the same object in the exact same situation and get the same results—it's very clear from quantum mechanics that that idea has to go. Physics tells us that there are no public physical objects" (Gefter, "The evolutionary argument," 2016).[57]

9. We only can say that something exists when we observe it.

If it is so that quantum events, and so eventually classical events as well, require observation, how is it that quantum events occurred before there was a participant observer—before there were humans, or animals, or life, or rocks, only primeval elements, or "quantum fluctuations?" We *do* conclude that a quantum collapse—a quantum event occurring at the subatomic level—*does* require an observer. That's what quantum theory maintains. *Quantum systems act as if a conscious observer-participant is required.* This raises a problem that is potentially unsolvable. Who or what observed the quantum events that occurred in the first moments of the big bang?! Martin Rees, British cosmologist and astrophysicist, sees it this way: "In the beginning there were only probabilities. The universe could only come into existence if someone

---

57 "Interestingly," continues Hoffman, "I can take two conscious agents and have them interact, and the mathematical structure of that interaction also satisfies the definition of a conscious agent. This mathematics is telling me something. I can take two minds, and they can generate a new, unified single mind.... I didn't expect that, the mathematics forced me to recognize this. It suggests that I can take separate observers, put them together and create new observers, and keep doing this ad infinitum. It's conscious agents all the way down" (Gefter, "The evolutionary argument," 2016). Here, Hoffman is speaking of what de Quincey refers to as "intersubjectivity," which we'll discuss more when we explore the notion of consciousness itself.

observed it. It does not matter that the observers turned up several billion years later. The universe exists because we are aware of it," (as quoted in Kitchen, 2017, p.178), (although I wonder, how could an observer "turn up" without being observed)?

Or perhaps, consciousness—observation—exists pervasively in the universe. In fact, the eminent physicist Freeman Dyson, almost incredulously has concluded, "It appears that mind, as manifested by the capacity to make choices, is to some extent inherent in every electron" (Frankenberry, 2008, p. 372).[58]

10. We subjectively participate in co-creating the universe.

The necessity for an observer at the quantum level leads to a conclusion that either *consciousness is primary (i.e., existed before time-space and matter/energy)* or *consciousness must have been an (emergent, nascent) property of the universe from the point of the big bang,* as we saw de Quincey asserted, (above in point nine). Consciousness is incipient in the universe in some form, resulting in an observation that reveals the universe at large. There doesn't seem to be another reasonable possibility.[59]

By virtue of our being conscious, by virtue of our being observers,

---

58 Dyson's full quotation, which may be interpreted in several ways is: "It is remarkable that mind enters into our awareness of nature on two separate levels. At the highest level, the level of human consciousness, our minds are somehow directly aware of the complicated flow of electrical and chemical patterns in our brains. At the lowest level, the level of single atoms and electrons, the mind of an observer is again involved in the description of events. Between lies the level of molecular biology, where mechanical models are adequate and mind appears to be irrelevant. But I, as a physicist, cannot help suspecting that there is a logical connection between the two ways in which mind appears in my universe. I cannot help thinking that our awareness of our own brains has something to do with the process which we call 'observation' in atomic physics. That is to say, I think our consciousness is not just a passive epiphenomenon carried along by the chemical events in our brains, but is an active agent forcing the molecular complexes to make choices between one quantum state and another. *In other words, mind is already inherent in every electron, and the processes of human consciousness differ only in degree but not in kind from the processes of choice between quantum states which we call 'chance' when they are made by electrons."* From *The Faith of Scientists: In Their Own Words* edited by Nancy K. Frankenberry, 2008. Princeton University Press, Princeton, New Jersey, p. 372. Dyson is an American theoretical physicist and mathematician, born in England, and is known for his work on nuclear reactors, solid state physics, ferromagnetism, astrophysics and biology quantum electrodynamics, and nuclear engineering.

59 Actually, there is another conjecture—the one that scientists today generally assume (with no proof), and then use to interpret data based on it. That is materialistic: that subjectivity, sentience, and consciousness somehow resulted out of objective, insentient, and unconscious matter. Materialism rejects the miraculous, yet somehow argues for the faith-inspired miracle that mind—in some way science hasn't begun to articulate—arises out of inert matter.

the universe emerges. Without our conscious observation, we must conclude nothing exists. With our conscious observation. we participate in *choosing* what will manifest. Yet, materialist science does not even attempt to address the phenomenon of choice in its assertion that all is physical! Indeed, "Nowhere in the reductionist worldview does one find an account of the emergence and reality of agency in the universe," observes Kauffman" (Kauffman, 2008, p. 11).[60] It seems then we need to broaden our definition of consciousness as (at least) an emerging, evolving, co-existent aspect of "material" reality. Agency—the "capacity, condition, or state of acting or of exerting power," as defined in Merriam-Webster, and based in a "capacity to know" (Narby, 2006)—does exist, in humans, as well as in fauna and even flora, as we will see, and further as Dyson suggested, exists "to some extent" all the way down to the electron. J. Conway and S. Kochen propose what they call "the free will theorem," which states that, "If any part of the universe has free will, it all does, but if any part doesn't, then none of it does—so either nothing is conscious or everything is" (Conway and Kochen, 2006).[61]

11. Observation is consciousness, and quantum systems require a conscious observer.

Further, since our universe is the outcome of the quantum expansion we call the big bang, and we—you and I—are outcomes of the big bang, *we are one quantum entity,* an interconnected wholeness.[62] John Bell's theorem seals the deal. He set about to prove that Einstein

---

60   Kauffman discusses agency and emergence: "You and I are agents; we act on our own behalf; we do things. Agency has emerged in evolution and cannot be deduced by physics. With agency come meaning and value. We are beyond reductionist nihilism with respect to values in a world of fact. Values exist for organisms, certainly for human organisms and higher animals, and perhaps far lower on the evolutionary scale. So the new scientific view of emergence brings with it a place for meaning, doing, and value." (Kauffman, 2008, p. 4).

61   Conway and Kochen write, "The world it presents us with is a fascinating one, in which fundamental particles are continually making their own decisions. No theory can predict exactly what these particles will do in the future for the very good reason that they may not yet have decided what this will be!" (Conway and Kochen, 2006). And Whitworth explains that in the quantum realism (virtual world) view, everything is conscious, there are no empty pixels and every electron is a "player."

62   Recall Brian Greene: "Even if you try to remove every atom and every particle, empty space is...alive with activity. Particles are continually popping in and out of existence, erupt out of nothingness, quickly annihilate each other and disappear."

was right: there cannot be "spooky action at a distance," there cannot be non-locality or entanglement or complementarity. But Bell's experiments, since often repeated by others, surprisingly led him to the opposite conclusion: there *is* spooky action—there is non-locality, and there is entanglement, and there is complementarity in the realm of the quantum—the realm that underlies all reality. Bell's theorem is considered to be one of the most significant experimental outcomes in the last 80 years. The unarguable conclusion that results? "In principle, any objects that have ever interacted are forever entangled, and therefore what happens to one influences the other," explain Rosenblum and Kuttner, (Rosenblum & Kuttner, 2011, p.188), and further, "Even events at the edge of the galaxy instantly influence what happens at the edge of your garden" (Rosenblum & Kuttner, 2011, pp. 173-174). All that exists, then is one conscious system, a unified conscious observer, entirely interconnected.

12. The universe is interconnected at the quantum level.

"There is, in principle, a universal connectedness whose meaning we have yet to understand," ratifies Marco Bischof, German theoretician (Rosenblum & Kuttner, 2011, p. 189). He observed, "Quantum mechanics has established the primacy of the inseparable whole. For this reason, the basis of the new biophysics must be the insight of the fundamental interconnectedness *within* the organism as well as *between* organisms, and that of the organism *with the environment*" (László, 2007, p. 49). This point is reiterated by biologists Elisabet Sahtouris, Lipton, and others, as we will explore ahead.

Walker contends that all time and space existed as "one thing, a quantum state," and more: "The inflationary theory together with the Big Bang theory let us trace everything back to the very beginning. That beginning is called the Planck era. And what do we find there? We find that the universe began as a primitive, pure, intense—a very intense—quantum state that existed in an infinitesimal point of space and time—a moment when all matter potentiality, all space, and all time existed as one thing, a quantum state...We have found that, above all else, that quantum states and mind are one and the same thing...We discover that in the beginning, there was the Quantum Mind, a first cause, itself time-independent and non-local, that created space-time and matter/energy" (Walker, 2000, p. 326).

Undeniably, we are one quantum entity. Everything is interconnected. It is *one* evolutionary process, *one* universal project, all, together, unfolding. That wholeness, the universe—given that we are alive, sentient, conscious, and self-reflectively aware, and given that "quantum states and mind are one and the same thing"—must *always* have been somehow alive, sentient, conscious, and self-reflectively aware, emerging to more diverse and competent forms as the universe unfolds, complexifies, and evolves itself. The entire process is a complex, continually unfolding dynamic that, quite obviously, must transcend and include the billiard ball-like mentality of reductionist physics.

So, to reiterate for a moment, quantum physics leads us to the following inferences, at the subatomic level absolutely, and at the macro level ultimately:

- Matter is not solid. It is activity—processes and events—which is the essence of being.
- We do not live in a world that always functions in a cause and effect way. The world consists of more than matter acting on matter.
- There is always uncertainty, and the world is indeterminate. We live in a world of probabilities, not predictability.
- Information and action can occur instantaneously across immense distances. It appears that an "entangled" particle *instantaneously* "knows" and reacts in a complementary fashion to the actions of its counterpart, even at unfathomable distances.
- Quantum events are uncaused, that is, self-caused. There is no cause outside of the quantum event itself that explains its action.
- Despite the illusion that we can predict events in advance of our observation, we can only determine the probability we might observe something from all available options.
- What happens in the world is affected by our observation. The

observer is not separate from the object being investigated… "The so-called observer is actually a *participator,* an integral part of the quantum system."

– We only can say that something exists when we observe it: otherwise, as the philosopher Ludwig Wittgenstein has said, "Whereof we cannot speak, thereof we should remain silent."- We subjectively participate in co-creating the universe. We each have our own unique consciousness. In quantum dynamics, there is no objectivity.

– Observation is consciousness, and quantum systems *require* a conscious observer.

– The universe is interconnected at the quantum level—the micro world that underlies the macro world of classical mechanics—unfolding, complexifying, and evolving itself as it expands.

The universe must somehow always have *been alive, sentient. conscious, and self-reflectively aware.* Because we have the capacity to know, obviously the capacity to know exists in the universe! These features occur as potentials that emerge and evolve out of the seeds of the big bang. They likely precede, but are, at least, co-existent with matter from the beginning.

Rather than coming from a world that is solely material, with cause and effect predictability, objectivity, separateness, and with consciousness emerging out of matter and limited to humans and some other life forms, *quantum theory tells us that the solidity of matter is an illusion; that a strict cause and effect world doesn't exist; that we can't reliably predict what will happen; that everything is somehow connected, and that our perceived boundaries "between" things—or "between" each other—are illusory; that reality is indeterminate and unpredictable, that we are participant/observers in creating the universe; that observation is required to know something exists; and that the universe must be alive, sentient, conscious, and possessing self-reflective awareness. It is and we are one quantum entity throughout.*

Fairly game changing.

A Holomovement of Unbroken Wholeness...
An Undivided Flowing Movement without Borders[63]

"Much of the apparent confusion around consciousness and quantum physics results from the efforts of scientific materialism to prevent their union, generally by denying that consciousness even exists or failing to grasp the deeper lessons of quantum evidence about their relationship. With a fresher perspective, some vexing problems in science become more approachable."

—Eben Alexander (Alexander, 2017, p. 52).

"It isn't that the world of appearances is wrong; it isn't that there aren't objects out there, at one level of reality. It's that if you penetrate through and look at the universe with a holographic system, you arrive at a different view, a different reality. And that other reality can explain things that hitherto remained inexplicable scientifically."

—Karl Pribram (as cited in Talbot, 1991, p. 11).

"When quantum theory implies that consciousness must exist, it tacitly shows that the content of the mind is the ultimate reality, and that only an active observation can confer shape and form to reality—from a dandelion in a meadow, to sun, wind, and rain."

—Robert Lanza (Lanza, 2009, p. 81).

The standard and most widely accepted interpretation of what the data from quantum physics tells us about reality, the Copenhagen interpretation, describes a universe that is action and process; becoming, emerging, and evolving in a self-caused, unpredictable, interconnecting, sentient, self-reflective, observing, creative, and participatory way.

In fact, quantum physics describes physical reality in such a new and unfamiliar way that it is difficult sometimes even to envision what it describes. It speaks of a world of black holes, changing time and a

---

63   This Bohm's basic tenet, the foundation of his thinking. Allan Combs & Mark Holland describe Bohm's thinking in *Synchronicity: Science, Myth, and The Trickster.* "The universe according to Bohm actually has two faces, or more precisely, two orders. One is the explicate order, corresponding to the physical world as we know it in day-to-day reality, the other a deeper, more fundamental order which Bohm calls the implicate order. The implicate order is the vast holomovement. We see only the surface of this movement

fullness of space, of fields and holograms, multiple dimensions of existence, and two dimensional worlds. Physicist David Bohm's hologram model of the Universe illustrates the challenge of description.

Bohm was one of many who tried to find a coherent theory in order to make sense of these mind-bending results of quantum research. He articulated a perspective that everything that exists is part of an "unbroken wholeness" which operates as an "undivided flowing movement without borders." His view of reality was that matter and life is a whole and integrated realm, which he called the Implicate Order. All of the objects in the universe, which he saw as the Explicate Order, are an "unfolding and enfolding" continuous process of movement, a primary and flawless whole.[64]

Before he died, he brilliantly articulated this theory of the universe, using the idea of a hologram in motion, a holomovement. At the time, (the 1960s-1980s) he was largely scoffed at and dismissed by the scientific community, but today, these ideas are taking center stage as a more persuasive description of our existence here.[65]

Brian Greene recalling Bohm's holomovement model, suggests that, as a result of recent work on the mathematics of black holes, it is thought now everything we experience as three dimensional *may actually be a projection from a far distant two dimensional surface that surrounds us*, which results in a hologram (Hickman, *The Elegant Universe*, 2003).

He uses the illustration of his wallet being pulled into a black hole.

---

as it presents or 'explicates' itself from moment to moment in time and space. What we see in the world — the explicate order — is no more than the surface of the implicate order as it unfolds. Time and space are themselves the modes or forms of the unfolding process. They are like the screen on the video game. The displays on the screen may seem to interact directly with each other but, in fact, their interaction merely reflects what the game computer is doing. The rules which govern the operation of the computer are, of course, different from those that govern the behavior of the figures displayed on the screen. Moreover, like the implicate order of Bohm's model, the computer might be capable of many operations that in no way apparent upon examination of the game itself as it progresses on the screen" (Combs and Holland, 1990). As we will see below. What sounds here like a computer as metaphor may indeed turn out to be more of an accurate descriptor of reality, as we'll discuss in some detail.

David Peat, in *Infinite Potential: The Life and Times of David Bohm,* in fact, describes this specifically: "According to Bohm, the ground of the cosmos is not elementary particles but pure process, a flowing movement of the whole (Peat, 1997).

64   Imagining what we want to make for dinner exists, as an example, in the Implicate Order of what's possible, a superposition. Actually making the meal occurs overtly in the Explicate Order, a collapse.

65   Bohm (who died in 1992) was one of the world's greatest quantum mechanical physicists and was deeply influenced by both the Indian philosopher Jiddu Krishnamurti and Albert Einstein.

The "information" about the wallet, he says, is stored (like information is stored in a computer) as two dimensions on the spherical periphery of the black hole, even as the actual 3-D wallet exists within the hole. (Imagine mirror-like icing on a donut and reflecting the "wallet" as it falls into the donut hole). He wonders whether this is an illusion of some kind—but muses that, as we have seen, the three-dimensional world of matter itself an illusion, after all (Hickman, *The Elegant Universe*, 2003).[66]

Consider that, as David Hawkins[67] observes, "in viewing a hologram, what you see depends completely on the position you view it from. So, what position then," he asks, "is 'reality'?" (Hawkins, 2014, p. 239). And how can there be an objective reality we can point to?[68]

Hawkins states clearly, "In fact, *this is a holographic universe.* Each point of view reflects a position that's defined by the viewer's unique level of consciousness…Each of these…goes 'back in time' to the

---

66   The holographic universe hypothesis predicts that, mathematically speaking, the universe needs just two dimensions (not three, as we perceive things to be) for the laws of physics and gravity to work as they should. Some physicists have suggested that "the reason we can't figure out what happens to stuff once it falls over the edge—or event horizon—and into a black hole, is because there is no 'inside'. Instead, everything that passes the edge gets stuck in the gravitational fluctuations on the surface." (Does an image penetrate a mirror and continue on its trajectory? Or does the process stop, and the image reflected back to the observer?)

    As reported on June 1, 2016 in *Science Alert,* entitled "The Case For Black Holes Being Nothing But Holograms Just Got Even Stronger," by Bec Crew, a team led by physicist Daniele Pranzetti from the Max Planck Institute for Theoretical Physics in Germany posed: "we now have a concrete model to show that *the 3D nature of black holes could just be an illusion.* All the information of a black hole can theoretically be contained on a two-dimensional surface, with no need for an actual 'hole' or inside." The article offers a visual for understanding: "…think of a black hole as a three-dimensional basketball hoop - the ring is the event horizon, and the net is the hole into which all matter falls and disappears. Push that net up into the ring to make it a flat, two-dimensional circle, and then imagine that all that metal and string is made of water. Now everything you measure in the ring can be applied to what's in the net" (Crew, "The Case," June 1, 2016). http://www.sciencealert.com/the-case-for-black-holes-being-nothing-but-holographic-images—just-got-stronger.

67   David R. Hawkins, M.D., Ph.D., is a psychiatrist and author. In the 1970's he co-authored with Nobel Laureate Linus Pauling, *Orthomolecular Psychiatry,* which revolutionized the field of psychiatry. He is also a consciousness researcher and mystic.

68   Adds Donald Hoffman, speaking in the YouTube video, "Simulation 2017: We Are Waking Up.": "You cannot, in the space around us, stick information into volumes…How much information could you actually stick inside, say the volume of that volleyball? … What's the maximum amount of information, in the ultimate, that you could stick inside that volume of that volleyball? Well, it turns out we have an answer. Stephen Hawking answered this and it's independent of the volume. The amount of information you can stick inside a volume of space does not at all depend on the volume. The volume is utterly irrelevant. You cannot use volume to store information—only surface area. The amount of information is proportional to the surface area. It's called the holographic principle. It's one of the deepest, most profound discoveries in physics in the last 30 years" Hoffman, 2017, *Simulation).*

original source of its existence, which is now" (Hawkins, 2014, p. 239). It is a world of many subjective realities.

Bohm believed also that "dividing up the universe into living and non-living things has no meaning. Animate and inanimate matter are inseparably interwoven, and life, too, is enfolded throughout the totality of the universe. Even a rock is in some way alive, says Bohm, for life and intelligence are present not only in all of matter, but in 'energy,' 'space,' 'time,' 'the fabric of the universe.' And everything else we abstract out of the holomovement and mistakenly view as separate things" (Talbot, 1991, p. 50).

What kind of fanciful, fantastic, and implausible world is science describing? Is this holomovement—with its swirling energy and its optical 3-D illusion of a projection from a two-dimensional distant perimeter, interconnected throughout, and alive throughout—possibly the reality of our existence?

What about space and time? Are they truly saying that they aren't constants? Fixed? The same for all of us? We know from Einstein that time and space are *not* constant, steady, unchanging realities: "…in the theory of relativity, there is no absolute time. In other words, each observer has (her or) his own measure of time" says Hawking (Hawking, 1988, p. 79). Even though we experience time as the same for all of us, we've learned that over great speeds and distances that time changes, and it is part and parcel of space and the objects contained in space. Space itself, according to relativity theory, is relative, curved, and finite. What time it is and "where we are" depends on who is doing the perceiving. Where we are in time and space determines what we experience, and that experience is not the same for anyone else. Our experiences of physical place and time are unique, thus relative and subjective, not absolute and objective. Says Brian Whitworth, "Space and time are convenient ideas for ordinary life but they don't explain the extraordinary world of physics" (Whitworth, 2014, *Quantum*).

Further, says de Quincey, scientists are conjecturing that the past and the future may perhaps exist as co-realities with our present.[69] There is a phrase that was made popular from the 1950's TV series *Science Fiction Theater* that perhaps portends: "Time is Just a Place"

---

69   See also, the PBS series *The Fabric of the Cosmos* (McMaster, *Fabric,* 2011).

(Arnold, "Time Is," 1955). Is the whole shebang akin to a gigantic cat's-eye marble, swirling and spinning and moving in formless imaginings, shape shifting, now at one time, now another, (and where)? Is the whole of existence contained in some two-dimensional yet spherical orb, reflecting inward and shaping our "reality?"

Or has science lost its collective mind?

ONE GIGANTIC PROCESS, A PROCESS OF BECOMING

If we believe that science has reliable and truthful things to say about the world we live in, we have to take to heart the current condition of quantum science—that the truth of it all is not something we've ever imagined, that it is even beyond the scope of our imagining, and yet, it manifests with an integrity and order far beyond randomness, chance, and grand scale chaos. Here are the words of a quantum physicist, a German philosopher and mathematician, a contemporary philosopher, a neurosurgeon, and an advocate of Darwinian evolution, who each experience our world in a far different way than traditional cultural understandings suggest:

- Erwin Schrödinger: "What we observe as material bodies and forces are nothing but shapes and variations in the structure of space. Particles are just *schaumkommen* (appearances). The world is given to me only once, not one existing and one perceived. Subject and object are only one. The barrier between them cannot be said to have broken down as a result of recent experience in the physical sciences, for this barrier does not exist" (Schrödinger, 1995).
- Gottfried Leibniz, (18th century German philosopher, mathematician, and logician): "Reality cannot be found except in One single source, because of the interconnection of all things with one another. ... I do not conceive of any reality at all as without genuine unity. ... I maintain also that substances, whether material or immaterial, cannot be conceived in their bare essence without any activity, activity being of the essence of substance in general" (Leibnitz, 1670).

- Arne Naess (Norwegian philosopher, founder of deep ecology): "Life is fundamentally one" (http://www.morning-earth.org/deep-ecology.htm).
- Eben Alexander, (neurosurgeon): "An overarching lessons from the study of modern physics is that the world is not as it appears. All of the basic components of the physical world (including our human brains and bodies)—the molecules, atoms, protons, neutrons, and photons of light, among others—are best viewed as vibrating strings of energy in higher dimensional space" (Alexander, 2017, p. 52).
- Thomas Huxley (defender and advocate for Darwin's theory of evolution by natural selection): "The different branches of science combine to demonstrate that the universe in its entirety can be regarded as one gigantic process, a process of becoming, of attaining new levels of existence and organization, which can properly be called a genesis or an evolution" (Teilhard de Chardin, 2008, p.13).

As we identify the single fabric, the oneness that is the cosmos, it seems that scientists, mathematicians and philosophers of varied ilk, somewhat ironically, are beginning to speak in a more unified voice. The physical world is not a world of material or matter; it is a world of emergent movement, continual activity, and evolving complexity, entirely connected as one entity.

The universe that was born out of the big bang is not purely material, and not just some initial, powerful expansion, and then silence. The energy and information released in the big bang continues today to swirl in space; to create planets, peonies, ponies, and people; and to continue to emerge—evolving new plots and new twists in the greatest page-turner of all time. The universe is emergence, movement, action, and evolution. And it is aware.

Because of the implications of quantum data, we know too that we participate in and shape the universe unfolding. We are even obliged to conclude that there must be subjective awareness, in some sense, in all of being.

Now we are seeing that *the cosmos consists in a grand and mysterious indissoluble vision, a union within which we are an inextricable part.* Science has taken on the study of matter, and has discovered its essence is well beyond material.

Chapter Four:

Quantum Dynamics— Interpretations and Implications

A Matter of Interpretation

"Everything we call real is made of things that cannot be regarded as real."

— Neils Bohr.[70]

Virtually all of the various conclusions that have been considered by physicists to try to make sense of quantum dynamics appear somewhat implausible and counterintuitive, and are clearly different than our daily lived experience. Since the discovery of the underlying quantum reality, physicists have struggled to find some alternative way of looking at the data so that it better matches what we are familiar with in the "material" world.

There is no argument, no debate, no challenge to the mathematical facts and technological realities that quantum theory describes. It

---

70   M. V. Echa, "On the Life and Legacy of Niels Bohr," in *ECHA & SCIENCE! The Official Website for Post-modern Physics*

doesn't jibe with the ways we see the world, but *it works* without exception, and in every case. It is the truth about our existence, despite the fact it flies in the face of the ways we experience our world. There are none who disagree. *It is not disputable that quantum dynamics describes the way the world works.*

What is disputable—and what falls outside of the realm of science and physics, and into the realms of philosophy and metaphysics—are questions such as why quantum mechanics is so; why it contradicts our everyday experience, yet is flawlessly applicable in our world; what it means for how we understand our world, and even for how we make sense of our existence, of who we really are. These are, plausibly, indeterminately complex and possibly unresolvable philosophical questions. "A word of warning:" says Alastair Rae, "this is an area of considerable controversy, where there are a number of alternative approaches, which means that our discussion is more philosophy than physics" (Rae, 2005, p. 176).

Interpretation always falls in the arena of philosophy, just as data always falls into the arena of science. It's essential to be clear about which discipline gets the microphone about which content. Obtained data may be indisputable: what the data mean, we will see, can be hotly disputed.

Tom Siegfried, in the Science News article, "Tom's Top 10 Interpretations of Quantum Mechanics," writes, "Dozens of interpretations of quantum mechanics have been developed over the years. Most of them attempt to address what happens when an observation or measurement is made on a quantum system. The mathematical formula known as the wave function (or state vector) describing the state of a system gets reset when a measurement is made, and the multiple possibilities that the math describes appear to 'collapse' into one tangible result. A quantum 'interpretation' tries to explain why this collapse happens—or whether it happens at all. And some interpretations concern themselves with whether the wave function itself is physically real or merely something mathematical" (Siegfried, *"Tom's,"* 2014).

"Quantum mechanics," Rosenblum and Kuttner caution, "shows that our reasonable, everyday worldview is fundamentally flawed. Interpretations of what the theory tells us offer different worldviews. But every one of them involves the mysterious intrusion of the conscious observer into the physical world" (Rosenblum & Kuttner, 2011, pp. 218-219).

Rosenblum and Kuttner use the expression "quantum enigma" to describe the conundrum of quantum theory: that it appears that we cannot say that anything at all exists until we observe it, that the act of observation creates the reality we know. So, observation—consciousness itself—is a prerequisite for the material world in which we find ourselves! This is the conclusion of Bohr and others, who, upon first encountering quantum, articulated the standard interpretation of quantum theory, the Copenhagen interpretation.

Over the twentieth and into the twenty-first centuries, physicists have posed other countering interpretations, though none have replaced the original Copenhagen interpretation as a consensus preference for how we might understand the physical world. Let's discuss Copenhagen and some of the other possible interpretations that have been proposed, as a way of exploring how scientists and science philosophy are constrained to proposing unconfirmed and unproven speculations as to what the data from quantum mechanics mean, and what they might suggest about the universe, life, mind, causality, and the reality of our existence here.

(A few of the interpretations described below may seem a bit opaque, with some scientific terminology that likely goes beyond most lay persons' ken. The evaluation of the interpretations is the relevant point. Feel free to skip this section if you find your eyes rolling up into the back of your head!)

Copenhagen

The accepted interpretation of the data from quantum physics, which we have introduced, is called the Copenhagen interpretation, because Niels Bohr, an originator of quantum theory, who received the Nobel Prize in Physics in 1922, was in Copenhagen when he described this way of looking at the data. He saw that an entity is whatever it is observed to be (either a wave or a particle), but that it cannot be assumed to have specific properties, *or even to exist,* until it is observed (measured). As the eminent physicist John Wheeler has put it, "No elementary phenomenon is a phenomenon until it is an observed (or registered) phenomenon" (Retrieved from Jones, "John Wheeler's" February 13, 2014).[71]

Each of us observes or measures from our own individual unique position in time and space, all in motion, as we've noted. That means that objective reality does not exist,[72] and it means that the other perplexing features of quantum theory we described above, do indeed reflect reality, and that our experiences and perceptions do not give us direct data about what is real.[73] What quantum theory concludes *simply doesn't support our sensory perceptions of a physical reality for our existence.*

Copenhagen also identifies the role of the measurer or observer as requisite for a "quantum collapse," that is, a resolution into one outcome vs. others. Because of this, the phenomenon of consciousness *must* be incorporated into our understanding of how reality works.[74] The Copenhagen interpretation defines observation as occurring whenever a *microscopic* object interacts with a large-scale *macroscopic* object. So, this viewpoint deals with classical mechanical macroscopic measuring instruments (whose results we observe), as well as we humans ourselves. It also deals with the microscopic quantum realms—of which macroscopic objects consist—according to the Schrödinger Equation (Rosenblum & Kuttner, 2011, p. 126). "Three principles—randomness of individual outcomes, alteration of state by measurement, and

---

71 "FORUM: From the Big Bang to the Big Crunch," North American AstroPhysical Observatory; Cosmic Search Volume 1 Number 4; Fall (Oct.- Dec. 1979, page 2, an "exclusive interview with John A. Wheeler was made by Mirjana R. Gearhart of Cosmic Search." At the time, it was believed the universe oscillated: It expands and then it contracts. Although this was found to not be true—the universe continues to expand—the point Wheeler made in that interview about observed phenomena remains a valid perspective to take.

72 This involves the principle called superposition that claims that while we do not know what the state of any object is, it is virtually in all possible states simultaneously, as long as we don't look to check. When we look, "collapse" occurs: all other options save the one we observe are no longer relevant to us. However, our unique observational point of view synchronizes to a sizeable degree with the viewpoint of others—as Hoffman (above) has found—and so *it seems as though* we will have some "objective," (conventional), shared agreement on what we are observing.

Eugene Wigner assumed that all humans, at least, possesses the status of "observer." But, as Rosenblum and Kuttner say (2011, p. 149), "All we know is that, someplace on the scale between big molecules and human awareness, there is this mysterious place of observation and collapse."

73 This conclusion will be a significant and primary focus of our discussion ahead, when we look at the phenomenon of consciousness itself.

74 Copenhagen acknowledges the indispensable conclusion that observation is required for a quantum particle to exist, but, like most current interpretations, does not attempt to seek understanding as to *why* and *how* this is so—and perhaps wisely, as these are questions more appropriately addressed by metaphysics and philosophy, not science.

our ability to calculate probability—underpin the conventional interpretation of quantum physics" states Rae (Rae, 2005, p. 181).

Many physicists, throughout the twentieth century, Planck and Einstein themselves among them, were very troubled by these conclusions, and the interpretation set the traditional Newtonian cause and effect physics on its head. (For one reason, for example, the question is begged: how could there be any deterministic, billiard-ball-like cause and effect if there is no objective truth or reality to begin with?!).

The relationship between macroscopic and microscopic dominions is, in quantum dynamics, both perplexing and significant. "Copenhagen treats the microscopic and macroscopic realms differently" note Rosenblum and Kuttner (Rosenblum & Kuttner, 2011, p. 127). Although quantum mechanics governs all that we observe in the physical world, we proceed *as if* there were two sets of laws: one for the macro world and another for the micro world. In 1932, confirming the value of that strategic tactic, John von Neumann demonstrated that "for all *practical* purposes we may consider macroscopic apparatus classically. This emphasizes that Bohr's separation of the microscopic and the macroscopic is only a very good approximation" (Rosenblum & Kuttner, 2011, p. 127). But the approximation is powerful. It allows us to play by Newtonian classical rules for big things and to play by quantum rules for tiny ones! Still, foundationally, that's illusory: the reality that ultimately governs macro and micro emanates from the subatomic, micro, reality of quantum mechanics.

Most physicists today, unlike Planck and Einstein, prefer to accept Copenhagen without further analysis, recognizing they are stepping into the arena of the philosophical, and out of the arena of the scientific, and seeing as well that science to date has provided no integrated answer.

Walker tells us, "Einstein, preoccupied with the desire to show that quantum physics was incomplete, did not see the philosophical system—the paradigm—on which he had based his physics was itself hopelessly incomplete. The world we know has conscious observers; the world Einstein described had particles moving along trajectories but no one to observe. Einstein did what many physicists have done: he assumed that consciousness and mind require no physical explanation" (Walker, 2000, p. 93).

At the moment of quantum collapse, Walker states, "we will see only one picture, one state, one condition. We know that state vector collapse must have already occurred or that it occurs at the time of this observation. As a result, investigators often speak of 'state vector collapse on observation.' But *observation* is just a euphemism for consciousness, for mind; this interpretation of quantum mechanics says that the system undergoes a state vector collapse because of our mind!

"...This effort to obtain an entirely practical interpretation of quantum mechanics—this Copenhagen interpretation—leads us to the incredible conclusion that mind, or consciousness, affects matter!" (Walker, 2000, p. 95).

Copenhagen describes a universe dependent on conscious observers, on mind, to generate reality. Without an observer, nothing exists. What does exist is merely a mathematical equation, (residing only in the human mind), that establishes the range of probabilities of *finding* a particular outcome—not the probability of it *being* there: "There is a crucial difference! The object is not there before you found it there... Somehow your looking," say Rosenblum and Kuttner, "*caused* it to be in a particular place...'observations' not only disturb what is measured, observations actually *produce* the measured result" (Rosenblum & Kuttner, 2011, pp. 81-82).

Although none of it made sense, experiment after experiment reached the same conclusion—and the proof is in the pudding, as we witness the technological magic that has resulted from quantum mechanics' applications. "In experiment after experiment, the quantum laws have always been right," notes Greene (McMaster, *Fabric*, 2011).[75] Lanza concurs: "All the entangled experiments of the past decades point increasingly toward confirming Copenhagen more than anything else" (Lanza, 2009, p. 58).

---

75 Perhaps not precisely relevant here, but useful to note as another layer of quantum complexity, (which de Quincey also noted, as we saw earlier), from *Quantum Enigma*, (Rosenblum & Kuttner, 2011, p. 153): A further implication resulting from a quantum collapse is that the act of observation creates, as well, a past that is consistent with the observation: This apparent fact might explain the past 13.7 billion years—human observation may have been required for the past to come to be, as Rees suggested: "It does not matter that the observers turned up several billion years later. The universe exists because we are aware of it." "Perhaps hardest to accept," say Rosenblum and Kuttner, "is the claim that your observation not only creates a present reality, but also creates a past appropriate to that reality."

Over the years, several other interpretations have been posited, although none have gained the broad acceptance that the Copenhagen interpretation enjoys. As we have noted, "Not all physicists and philosophers are content with positivism [that positive knowledge is based on natural phenomena and their properties and relations as verified by empirical science][76] and considerable effort has been made to develop alternative interpretations that would overcome this problem. All of these have their followers, although none has been able to command the support that would be necessary to replace the Copenhagen interpretation as the consensus view of the scientific community" notes Rae (Rae, 2005, pp. 190-191).[77]

Other more recent interpretations, which are all built on the Copenhagen interpretation, include *Relational Quantum Mechanics (RQM), QBism,* and *Consistent Histories.*

Relational Quantum Mechanics

RQM holds that the state of the quantum system[78] is the *relationship* between the observer/measurer and the system, and that the specifics of the observation hinge on the reference frame. The "state" is not the observed system itself, but the *relationship* between the system and observers, (because any observer is by definition a part of the system, not standing outside of it). Different observers may give different accounts of the same events—but there are no *external* observers. All observers are part of the relationship, the interaction, the system itself. Relational Quantum Mechanics stresses that the physical world can

---

76   Merriam Webster.

77   In 2011 a poll, *A Snapshot of Foundational Attitudes Toward Quantum Mechanics,* by Maximilian Schlosshauer, Johannes Kofler, and Anton Zeilinger, the authors write in their abstract: "Foundational investigations in quantum mechanics, both experimental and theoretical, gave birth to the field of quantum information science. Nevertheless, the foundations of quantum mechanics themselves remain hotly debated in the scientific community, and no consensus on essential questions has been reached. Here, we present the results of a poll carried out among 33 participants of a conference on the foundations of quantum mechanics. The participants completed a questionnaire containing 16 multiple-choice questions probing opinions on quantum foundational issues. Participants included physicists, philosophers, and mathematicians." The Copenhagen interpretation was most preferred, at 42%, followed by informational interpretations (RQM, QBism), at 24%. In third place was the Many Worlds interpretation (18%). http://arxiv.org/pdf/1301.1069v1.pdf).

78   Called the wave function.

be considered only in terms of relationships between systems, because all facts about the world are explicitly gleaned from *interactions* (Rae, 2005). (This interpretation resonates with the notion of relationship as primary in experience, and so defines who we are, who the observer is, and the "reality" of our collective perception. We will also see that an understanding of consciousness itself must include an appreciation of its intersubjective—relational—aspect). Other than this specific, it mostly aligns with Copenhagen from both scientific and philosophical vantage points.

Quantum Bayesianism

Closely related philosophically to RQM, Quantum Bayesianism (QBism), is also a variation of the Copenhagen interpretation of quantum mechanics. QBism takes probabilities to be personal judgments of the observers who are measuring quantum mechanics. Additionally, it defines a measurement to be any action that the observer/measurer takes to elicit a response from the world. Further, language is the only means by which different agents can attempt to compare their internal experiences, with the result that all we can mutually know is *limited by a given language's conceptual and contextual development.* In effect, we have two or more subjective observers trying to communicate with and understand each other by means of a limited and imperfect medium.

"In a sea of interpretations of quantum weirdness, QBism swims alone. The traditional 'Copenhagen interpretation' treats the observer as somehow standing outside of nature, imbued with mysterious wave-function-collapsing powers, governed by laws of physics that are different from those that govern what's being observed. That's all well and good until a second observer comes along to observe the first observer,'" notes Christopher Fuchs, developer and a main proponent of QBism (Gefter, "A private," June 4, 2015).

The Copenhagen interpretation we've been discussing, along with the Many Worlds and Hidden Variable interpretations we'll discuss below, "all have something in common," Fuchs argues. "They treat the quantum state of a system as a description of *an objective reality shared by multiple observers*" (Gefter, "A private," June 4, 2015). It seems perhaps that QBism is a refinement of RQM, and that the two, taken together, further elucidate Copenhagen and reframe it through a subjective lens rather than an objective one. "QBism treats the wave function as a description of *a single observer's subjective knowledge* [emphasis added]," says Fuchs, (Gefter, "A private," June 4, 2015), akin to RQM, which describes the *subjective relationship* of observers and system.[79]

One of the stark difficulties with all interpretations inspired by Copenhagen as well as of Copenhagen itself is the lack of any serious and considered exploration or articulation of *the phenomenon of consciousness* that underlies any observation and all observers: *What,* after all is consciousness, and *how*, in these interpretations, does consciousness interact with the physical world such that any given material outcome results? Copenhagen and virtually all other interpretations, as we have noted, remain silent on this essential aspect of quantum mechanics. However, in order to begin to understand the influence of consciousness as a scientifically sound aspect of our world, first *acknowledging and describing consciousness* would be required for science to truly weigh in on, specifically, how it actually does interact with the material aspects of our world.

Many Worlds

A recent and provocative alternative that has been proposed, and to which a number of physicists subscribe—Hawking, Greene, and distinguished physicist Richard Feynman among them—is referred to as the Multiverse[80] or Many Worlds interpretation. (MWI, the Everett interpretation, the theory of the universal wave function, the relative state formulation, many-universes interpretation, parallel universe, and many-worlds are other names for this interpretation). It holds that *every conceivable outcome to an event will occur*, each in its own universe,

---

79 QBism seems to describe a solipsistic viewpoint, that the self is all that can be known to exist, which can never be proven untrue, and may have merit, but which most folks dismiss because it doesn't match our lived reality—not that much of what goes on in quantum matches our daily lived reality! Lanza actually argues for this point of view in *Biocentrism,* (Lanza, 2009). Along with all those of the persuasion that reality occurs only in the mind, we too will argue the merits of this perspective.

80 *Not the same* as László's metaverse!

and that we, in our known universe, experience one of those outcomes, while each of the other (infinite) possibilities actually occurs as well, in inaccessible (to us) parallel universes.

Because this interpretation is being given serious consideration in scientific circles these days, I offer a perhaps too extensive critique of its implications. I apologize in advance, should this seem the case for any reader. But as we will see, I'm far from the only critic!

Sarah Scoles, a writer at Smithsonian.com, describes the concept this way:

> The universe began as a Big Bang and almost immediately began to expand faster than the speed of light in a growth spurt called "inflation." This sudden stretching smoothed out the cosmos, smearing matter and radiation equally across it like ketchup and mustard on a hamburger bun.
>
> That expansion stopped after just a fraction of a second. But according to an idea called the "inflationary multiverse," it continues—just not in our universe where we could see it. And as it does, it spawns other universes. And even when it stops in those spaces, it continues in still others. This "eternal inflation" would have created an infinite number of other universes.
>
> Together, these cosmic islands form what scientists call a "multiverse." On each of these islands, the physical fundamentals of that universe—like the charges and masses of electrons and protons and the way space expands—could be different.
>
> Cosmologists mostly study this inflationary version of the multiverse, but the strange scenario can takes other forms, as well. Imagine, for example, that the cosmos is infinite. Then the part of it that we can see—the visible universe—is just one of an uncountable number of other, same-sized universes that add together to make a multiverse. Another version, called the "Many Worlds Interpretation," comes from quantum

mechanics. Here, every time a physical particle, such as an electron, has multiple options, it takes all of them—each in a different, newly spawned universe.

But all of those other universes might be beyond our scientific reach. A universe contains, by definition, all of the stuff anyone inside can see, detect or probe. And because the multiverse is unreachable, physically and philosophically, astronomers may not be able to find out—for sure—if it exists at all.

Determining whether or not we live on one of many islands, though, isn't just a quest for pure knowledge about the nature of the cosmos. If the multiverse exists, the life-hosting capability of our particular universe isn't such a mystery: An infinite number of less hospitable universes also exist. The composition of ours, then, would just be a happy coincidence (Scoles, "Think big," April 9, 2016).

This overarching notion that a physical reality literally exists for every conceivable micro-moment and micro-entity has resulted, due to its seeming simplicity and absurdity, in stark mockery in some circles. For example, Whitworth described it this way: "In 1957, Everett devised the fairytale…called many-worlds theory—that every quantum choice spawns an entire new universe" (Whitworth, 2014, *Quantum*). He is not alone in this opinion.

Philip Ball notes, "It is grounded in a half-baked philosophical argument about a preference to simplify the axioms."[81] Alastair Rae observes: "It turns out that one way to avoid this problem [of observation] is to ignore it" (Rae, 2005, p.197). Walker also comments that a "… universe of universes would be piling up at rates that transcend all

---

81   In "Too many worlds: Nobody knows what happens inside quantum experiments. So why are some so keen to believe in parallel universes?" by Philip Ball, in the digital magazine *Aeon*, (https://aeon. co/essays/is-the-many-worlds-hypothesis-just-a-fantasy), the author sums up his assessment of MWI thus: If the MWI were supported by some sound science, we would have to deal with it …But it is not. It is grounded in a half-baked philosophical argument about a preference to simplify the axioms. Until Many Worlders can take seriously the philosophical implications of their vision, it's not clear why their colleagues, or the rest of us, should demur from the judgment of the philosopher of science Robert Crease that the MWI is 'one of the most implausible and unrealistic ideas in the history of science'" (Ball, February 17, 2015, "Too many").

concepts of infinitude" (Walker, 2000, p. 107). Physicists seem to love it or to hate it!

Why some physicists prefer this interpretation is that it can seem to take the questions of measurement, observation, and consciousness itself off the table. In this interpretation, there are now an infinite number of "yous" in an infinite number of scenarios in infinitely minute units of time and space and matter, each with its own universe. With each action in any of these universes, presumably, another round of infinite universes are born, over and over, dizzyingly, *ad nauseum.* None of these alleged universes have any contact with each other, so I will never meet another version of myself. Therefore, what is real is not dependent on conscious observation.

In the *Scientific American* article entitled "Does the multiverse really exist? Proof of parallel universes radically different from our own may still lie beyond the domain of science," George Ellis writes: "For a cosmologist, the basic problem with all multiverse theories is the presence of a cosmic visual horizon. The horizon is the limit to how far away we can see, because signals traveling toward us at the speed of light (which is finite) have not had time since the beginning of the universe to reach us from farther out. All the parallel universes lie outside our horizon and remain our capacity to see, now or ever, no matter how technology evolves. In fact, they are too far away to have had any influence on our universe whatsoever. That is why none of the claims made by multiverse enthusiasts can ever be substantiated...

"We just do not know what actually happens, for we have no information about these regions, and we never will" (Ellis, "Does the multiverse," December, 2015).[82]

Further, Rae notes that, "Detailed quantum calculations show that it would be impossible for the 'me' that is one half of the superposition to be aware of the 'other me' in the other. This means that the whole system has 'branched' and there is no way that an observer on a branch...can ever know anything about the existence of the (other branched 'me') ... (P)rocesses are happening all the time in the physical

---

82 According to Wikipedia, Ellis "co-authored *The Large Scale Structure of Space-Time* with University of Cambridge physicist Stephen Hawking, published in 1973, and is considered one of the world's leading theorists in cosmology." http://www.public.asu.edu/~atpcs/atpcs/Misc/DoesTheMultiverseReallyExist.pdf

world, which would by now have created an unimaginably large number of branches." (Rae, 2005, p. 198).

Rae tells the reader that "the many-worlds theory has been described as 'cheap on postulates, but expensive on universes,'" and argues that "there is a fifty percent probability of (a particle) emerging... which implies that there is a fifty percent probability of it *not* doing so. But this is inconsistent with the fact that in a many-world scenario everything happens."

Rolston weighs in as well on the Many Worlds kinds of interpretations: "...these are complex explanations indeed—to invent myriads of other worlds existing sequentially or simultaneously with ours, in order to explain how this one can be a random one from an ensemble of universes, and so a little less surprising in its anthropic features. There is of course little or no scientific evidence that other universes exist. They are mostly scientific conjectures, if indeed these are scientific hypotheses of the familiar kind at all, since the existence of other, non-observable universes with differing constraints and laws is not directly testable, only marginally so by extrapolation from what we may know about this one" (Rolston, 2010, p. 29).

Rolston further cites Martin Gardner who said, "The stark truth is that there is not the slightest thread of reliable evidence that there is any universe other than the one we are in. No multiverse theory has so far provided prediction that can be tested" (Gardner, 2003, p. 9, as referenced in Rolston, 2010, p. 30). MWI, for many experts, it seems, has a credibility problem!

I, too, personally find this interpretation to be incredulous, untenable, nonsensical, and inconsistent. According to quantum theory, an observation is still required for a collapse into this universe's reality, and the features of observation (and so, consciousness), remain unaddressed and even unacknowledged. What is the actual observation that results in the creation of any alternate universe? Quantum data insists that observation is necessary for anything to exist, crazy as that sounds. So who is making the observation, and from what universe are they? Are there, then, multiple or infinite observers?

What sounds crazier is the purely capricious notion of "many worlds." What are the "units" of time or of space or matter that result in new, unique universes? One second, a millisecond, two hours—and experienced by whom? What spatial units morph to new universes: a plant, an animal, a planet? A cell, a molecules, an atom…or an event or circumstance?

We appear to live in a world of meaning. The entire scientific enterprise is premised on the quest for deeper understanding—deeper meaning. The infinities of MWI seemingly result in "meaning" that is arbitrary, pointless, and nonsensical.

Additionally, as quantum mechanics tells us, everything is somehow interconnected, everything is entangled, and if so, then how are all these by definition separate universes connected in one underlying quantum mechanical unity? How do they instantaneously influence each other, especially over a period of time or non-locally over unqualifiable space? *Do* they? Is each "universe" in relationship with the others, and if not, how could that be possible given our current understandings of the way our reality is organized under the quantum mechanical rubric? If they are interconnected, they aren't separate universes. If they are not interconnected, they defy the conclusions of a unified field of quantum reality. None of these subtleties appear to be addressed, articulated, or answered by MWI proponents. And there is only limited, highly speculative and untestable "data" to argue such a case—so the net is that we must conclude that MWI is purely whimsical philosophical speculation noteworthy because it is apparently not at all based in empirical science.

These questions pose problems that conflict with what we have learned to date about the universe—questions of observer, observation, consciousness, nonlocality, entanglement, complementarity, superposition and collapse, absurd infinite infinities empty of meaning, macro and micro interrelationships, interconnection, notion of the term "universe" itself, and an absence of evidence and empirical data, among others. Hawking has said, "Any physical theory is always provisional, in the sense that it is only a hypothesis: you can never prove it…On the other hand, you can disprove a theory by finding even a single observation that disagrees with the predictions of the theory. As philosopher of science Karl Popper has emphasized, a good theory is characterized by the fact that it makes a number of predictions that could in principle be disproved or falsified *by observation*" [italics added] (Hawking & Mlodinow, 2005, p. 14). What predictions does the multiverse interpretation make that can be shown to have any influence on the world we observe? Reasonably, we could contend that, indeed, this interpretation of the quantum data is no more substantive than the breath of angels. If this is the solution to the "quantum enigma," it seems modern science has become a cartoonish caricature of itself!

If we consider all that we know or could know as part of our universe, our cosmos, our bubble, then anything we might perceive or conclude is, de facto, a part of that bubble. We will never have the capacity to know of anything beyond or outside of our bubble. It can only be conjecture or fantasy. We'll have more to say on this in Chapter 6 and beyond.

Hidden Variables

Another influential quantum interpretation is the Hidden Variables interpretation, otherwise known as the DeBroglie-Bohm Interpretation (DBB), Bohmian mechanics, the Bohm interpretation, the causal interpretation, and the pilot wave theory. This interpretation says that quantum theory does not fully explain the complete state of the physical system, so is itself not complete.

This interpretation refers to "hidden variables," meaning that quantum objects do have physical attributes, even though we cannot see them. For example, position and velocity would theoretically be able to be measured, voiding Heisenberg's Uncertainty Principle: there would be no uncertainty. Bohm additionally considered the "Implicate Order" as necessary to understanding the complete system, and this was not accessible to physical—or "Explicate Order"—study.

Early on, Einstein himself was a proponent of this interpretation, (legendarily he said, "God does not play dice with the universe"), because he strongly believed in a deterministic, cause and effect world. "The theory is fully deterministic; if you know the initial state of a system, and you've got the wave function, you can calculate where each particle will end up," comments Dan Falk in a *Quanta* magazine article

(Falk, May 16, 2016, "New support).[83] Bohm's interpretation, however, requires a "hidden variable" of nonlocality, with faster than light speeds, which is inconsistent with a deterministic explanation.

Lanza's viewpoint is that "... Experiments of the past decade truly seem to prove that Einstein's insistence on 'locality'—meaning that nothing can influence anything else at superluminal speeds—is wrong. Rather, the entities we observe are floating in a field—a field of mind... – that is not limited by the external *space-time* Einstein theorized a century ago" (Lanza, 2009, p. 53).

Eventually, Einstein rejected the Bohmian view as being too simplistic. Still, this interpretation is considered the most straightforward and compatible with our lived experience, even though there have been consequential and significant challenges.[84]

Specifically, if this interpretation was valid, nonlocality would be required to be understood as one of those hidden variables. However, as we noted, nonlocality violates Einstein's relativity theory which creates an unaddressed inconsistency. A deterministic, classic-like approach rules out nonlocality (i.e. entanglement, Einstein's "spooky action at a distance"), decidedly nondeterministic features. How can a deterministic model have as an essential aspect a feature that is itself nondeterministic?

Ultimately, a death knell was delivered to the deterministic Hidden Variables interpretation of quantum mechanics when John Stewart Bell, a Northern Irish physicist, *and himself an advocate of this interpretation,*[85] in 1964 was able to determine unequivocally that no hidden variable model could ever reproduce all of the predictions of quantum physics, a finding that was later indisputably ratified in several rigid experiments using advanced technology, by both John Clauser and Alain Aspect (Rae, 2005, pp. 191-197).

---

83 The article reports on ongoing experimentation suggesting that both the position and velocity of particle may be knowable, but acknowledges that "even for those who embrace the Bohmian view, with its clearly defined particles moving along precise paths, questions remain. Topping the list is an apparent tension with special relativity, which prohibits faster-than-light communication...Many physicists feel that more clarification is needed, especially given the prominent role of non-locality in the Bohmian view. The apparent dependence of what happens *here* on what may be happening *there* cries out for an explanation."

84 By Bohr (the Bohr Einstein debate), by the EPR controversy, and by Bell's theorem, among others.

85 Bell died abruptly in 1990 in Geneva of a cerebral hemorrhage. He was awarded the Nobel Prize posthumously.

Consistent/Decoherent Histories

The Consistent Histories and Decoherent Histories interpretations generalize from the Copenhagen interpretation, but are more elaborate, more precise, and include classical physics, (FAPP), resulting in an "updated" version of Copenhagen. Measurement no longer plays a key role, and measurements are regarded in the same way as all other physical phenomena, playing no special part in interpretation—so there is no observer/measurer problem.[86] Quantum mechanics, in these interpretations, readily aligns with Einstein's special relativity, so that nothing moves faster than light, as he had asserted. There is no "spooky action at a distance."

If a number of histories[87] is consistent, one is able to establish probabilities consistently. Regarding this interpretation, the *Stanford Encyclopedia of Philosophy*[88] notes, "Classical mechanics emerges as a useful approximation to the more fundamental quantum mechanics," yet, "(a)n important philosophical implication is the lack of a single universally-true state of affairs at each instant of time," which results in the varied subjective viewpoints of quantum theory, verses any objective classical possibility.

There are several problems with Copenhagen that the Consistent/Decoherent Histories interpretations can clear up, using some slightly different assumptions to start with and introducing additional concepts and relationships, all at a highly technical level. The Consistent Histories interpretation considers classical mechanics as a mere rough calculation within quantum mechanics, and so the quantum math can be used to compute probabilities for both macro and micro phenomena. In this interpretation, probabilities are based on physical states within the system, and not on the results of measurements: probabilities are not real physical states. A wave function or mathematical statement for calculating probabilities of sequences of events, or histories

---

86   Not acknowledging however that interpretation is itself not a physical phenomenon.
87   "A *history* is a sequence of quantum events (i.e., wave functions) at successive times," notes the Quantum Physics at Carnegie-Mellon University webpage, http://quantum.phys.cmu.edu/CHS/histories.html. According to the Quantum Physics at Carnegie-Mellon University webpage, http://quantum.phys.cmu.edu/CHS/histories.html,
88   The Stanford Encyclopedia of Philosophy (1995) is an online encyclopedia of philosophy with peer-reviewed publication of original papers in philosophy, maintained by Stanford University.

unfolding over time does not describe an actual physical state, but is simply a mathematical construct. The probabilities, then, that describe *the physical state* of a quantum system tell a consistent story about the quantum system as a whole.[89]

"(The) history approach, although it was initially independent of the Copenhagen approach, is in some sense a more elaborate version of it," according to theoretical physicist Roland Omnès. In combination with understandings that come out of the Decoherent Histories interpretation, it is accepted that unalterable macroscopic events result in all histories being automatically consistent (Siegfried, "Tom's," 2014).

Related to the Consistent/Decoherent Histories interpretations, *Quantum Darwinism* describes a process by which possible quantum realities are eradicated when a system interacts with its environment, a dynamic referred to as decoherence. Imagine that in a given environment, air molecules or photons bound off an object, quickly leaving only one position remaining that is consistent with the data chronicled in the environment. These "natural interactions produce a sort of 'natural selection' of properties that are recorded in the environment," says Siegfried (Siegfried, "Tom's," 2014).

The Decoherent Histories interpretation specifically reasons that the whole universe can be considered a quantum system with no external environment. So, decoherence occurs within the universe, and generates what are called "quasiclassical domains," where in quantum events can be treated as though they were classical mechanical (Siegfried, "Tom's," 2014).

Rosenblum and Kuttner identify an important issue with these interpretations: "Since no observer, conscious or otherwise, need be mentioned, some argue that this resolves the observer problem. Others see a fundamental non-sequitur in that argument. Those classical-*like* probabilities are still probabilities of what will be *observed*. They are *not* true classic probabilities of something that actually *exists*. Decoherence is then merely a FAPP solution to the quantum enigma" (Rosenblum and Kuttner, 2011, pp. 209-210).

Siegfried notes that, "after all, *people* are involved in using the

---

89   Retrieved from the Quantum Physics at Carnegie-Mellon University webpage, http://quantum.phys.cmu.edu/CHS/histories.html.

quantum math to compute probabilities and make measurements. Further, Gell-Mann and Hartle (originators of the Decoherent Histories interpretation) are undeniably correct that people exist within a closed-system universe and should be governed by the same physical laws as everything else…." (Siegfried, "Gell-Mann" 2014). The phenomenon of observation simply cannot be eliminated as a key ingredient of measurement.

I found myself confused with the idea of eliminating measurement or observation when one considers "the quantum system with no external environment," as it seems inconsistent. It is still observation that is required for a quantum system to exist. Also, "mathematical constructs" essentially fall into the category of measurement/observation. Mathematical quantum physicist Henry Stapp's consideration of Whitehead's comments below seem to address my questions here, and disavow validity for these interpretations, it appears:

"In any case, similar issues about the nature of reality occupied deep thinkers long before quantum physics came along. In a rough way, the array of competing quantum interpretations today parallels the old question of whether reality is an objective feature of the world or something in the mind, a question that mathematician-philosopher-theologian Alfred North Whitehead addressed: interactions of objects in space and time cannot be separated from perception of them, Whitehead pointed out. 'You cannot tear any one of them out of its context,' he wrote. 'Yet each of them within its context has all the reality that attaches to the whole complex.' The universe evolves during the continuous process of perceptions being unified with what's being perceived…'Thus nature is a structure of evolving processes,' Whitehead asserted. 'The reality is the process.'" (Stapp, 2001, "Quantum Theory").

Consciousness Causes Collapse

Consciousness Causes Collapse, otherwise known as the von-Neumann-Wigner interpretation, defines the quantum system and its relationship with humans (and perhaps, for some, animals, and possibly even plants or rocks or subatomic particles) with respect to consciousness itself. Henry Stapp argues, "From the view of the mathematics of quantum theory it makes no sense to treat a measuring device as intrinsically different from the collection of atomic constituents that make it up. A device is just another part of the physical universe... Moreover, the conscious thoughts of a human observer ought to be causally connected *most directly and immediately* to what is happening in his [or her] brain, not what is happening out at some measuring device...Our bodies and brains thus become...parts of the quantum mechanically described physical universe. Treating the universe in this way provides a conceptually simple and logically coherent theoretical foundation" (Stapp, 2001, "Von Neumann's").

Considering the material nature of reality presumed to be true by many physicists, this interpretation is often dismissed, because it is thought to be akin to Descartes' dualism. It also raises but does not address the question of which "things are capable of collapsing the wave function into one outcome." For others, such as Penrose, the question of the physical world's existence from the point of the big bang forward cannot be satisfactorily answered unless the physical world requires consciousness for existence. Still others are fearful of a solipsistic (that knowledge outside of one's own mind might not exist *outside of* one's own mind) or panpsychic (that consciousness is a universal and elemental feature of existence) point of view.

For me, the categorical dismissal of the role of consciousness, and at times even its very existence not being acknowledged as a component of reality that might be taken into account—despite the incontestable data from quantum experiments—represents a bias guaranteed to derail any truly viable answers to the questions that quantum data present. The Consciousness Causes Collapse interpretation doesn't answer the "quantum enigma," but it raises some important aspects—macro verves micro processes, observation and who possesses consciousness, solipsism and panpsychism, the relationship between "physicality" and consciousness, and others—that compel us to not settle for a simplistic, purely material understanding.

Other Interpretations

There are various other interpretations of the data from quantum physics. *Stochastic Mechanics, Transactional interpretations, Objective Collapse* (such as the *Ghirardi-Rimini-Weber theory (GRW); and the Penrose interpretation), the Many Minds interpretation, the Ithaca interpretation, the Quantum Information interpretation, Modal interpretations, Time Symmetric Theories, Branching Space-time Theories, Quantum Logic,* and others, not to be marginalized, all include important insights and questions—far more questions than answers!—but the reader can see from the partial survey above that the challenges and complexity involved in understanding a theory that, ironically, has so non-controversially and comprehensively transformed our culture are immense, and remain open to a wide range of potential perspectives that a discerning non-physicist might prefer.

Each of the possible interpretations we've identified have merit—but each has significant questions, and none have been verified. "In the absence of empirical differences between the interpretations, it is only natural to conclude that one's decision which interpretation to adopt will be influenced by personal preferences and beliefs," conclude the authors of the afore-mentioned 2011 poll, which leads us to *subjective conjecture* rather than *objective data,* and *philosophy* not *documented science.* There is no significant plurality within the scientific community *at all,* it seems, for how we should make sense of quantum dynamic reality, and there are no consequential data that should lead us to one interpretation over another. Given that, we are afforded the luxury of developing or adopting our own preferred (metaphysical) interpretation of the dynamic data of quantum mechanics. We may even question the primary assumption in these interpretations: that matter is primary, as we will see.

We have a number of competing interpretations and theories, and these are almost entirely based in an assumption of the primacy of matter: It's a physical world we live in, and we ourselves are physical. It's what we relate to, and it's functioning is based in natural laws that the discipline of physics, primarily, has discovered for us and explained to us. Material first, and then, inexplicably from the material, life, mind, and self-reflective awareness.

But what if this assumption is a false one? What if an assumption of physical realism doesn't reflect the truth of existence after all? That's a difficult hurdle, because of the obvious materialistic bias in our culture, and since, as we have seen, the hegemony of religion continues to cede more and more authority to science, to the extent that even the dualism of Descartes is dismissed:

"Dualism let the spirituality of religion coexist with the physicality of science, but divided scientists into atheists who believed only in the physical world, theists who believed in a world beyond it as well, and agnostics who didn't know," comments Whitworth. "Today, dualism seems increasingly an illogical kludge...Dualism is currently in retreat before the monism that there is only one real world. Scientists naturally think that if there is only one reality, let it be the physical one we study. The rise of physical monism inevitably conflicts with idealistic dualism." (Whitworth, 2014, *Quantum*). Physicalists prefer physical interpretations, and idealists prefer either dualistic (mind and matter coexist, but independently) or monistic (the primacy of consciousness) interpretations.

When we described in Chapter Two the astonishing inferences that quantum dynamics theory leaves us with, we saw that, in effect, for quantum theory to be true, physical realism, and even dualism, must not be—physicality is illusory and virtual, not "real." How can we move forward, then, and find a perspective that integrates what we have learned, and has the added value of not only functioning like we see quantum dynamics does, but also feels a bit sane in our lived physical experience? The conjecture we discuss in Chapter 6, I am hoping, might bring us a bit closer to that aim; but first, let's focus our attention on the idea of consciousness and what it might involve—especially since we are claiming that it is the true nature of the cosmos!

One can see from the reality of the limits of our understanding of quantum (that there is no physicality at the tiniest level of the subatomic world), that what we interpret as physical is based in non-physical. Yet we continue to assume a physical world, and our physics persists in interpreting quantum data as representing something physical when it can only be that our existence here is not at all physical, but mental— illusory, a virtual world! It is a world solely based in our observation, based in our consciousness.

Chapter Five:

Consciousness—The Ultimate Nature of the Cosmos

The Ultimate Nature of the Universe Is More Consciousness than Matter

"I regard consciousness as fundamental. I regard matter as derivative from consciousness. We cannot get behind consciousness. Everything that we talk about, everything that we regard as existing, postulates consciousness."

—Max Planck (Sullivan, January 25th, 1931," Interviews with").

"You have never witnessed anything other than the inside of your own consciousness."

Eben Alexander (Alexander, 2017, p. 53).

"The stream of knowledge is heading towards a non-mechanical reality; the Universe begins to look more like a great thought than like a great machine. Mind no longer appears to be an accidental intruder into the realm of matter... we ought rather hail it as the creator and governor of the realm of matter."

—Sir James Jeans, *The Mysterious Universe* (Jeans, 2010, p. 137).

"The more detailed one's analysis of the structure of 'out there' is, the more one discovers that what one is examining is, in fact...the nature of consciousness in here. There's nothing 'out there' other than consciousness itself," says David Hawkins (Hawkins, 2014, p. 245). Deepak Chopra concurs: "Perception is the world: the world is perception" (Chopra, 2004, p. 27).

Inadvertently, by the empirical quest to understand "out there," modern science is confronted with "in here." And even though science has yet to take on in a meaningful way the existence of consciousness itself—the very means by which we know anything—we nevertheless have become a self-reflectively conscious species.[90] We have evolved beyond any other life form we have detected to become *conscious, intelligent, reflective, perceptive, agentic,* and *relational* beings.

"The paradox facing science...is that for science to exist at all it requires experiencing beings who can *know,"* asserts de Quincey. "Yet the subjective experience *of any knower* is precisely what modern science cannot account for. If science is ever to become a *comprehensive* exploration of the cosmos, then, clearly, it must find a way to study consciousness. To overcome this glaring omission, science must begin by acknowledging that the cosmos possesses *consciousness that knows* and gives rise to science in the first place. Otherwise, by default, science will remain incomplete" (de Quincey, 2005, p. 208).

"To a growing number of scientists and philosophers it is beginning to look suspiciously as if the ultimate nature of the universe is much more akin to what we call 'consciousness' than matter," says de Quincey (de Quincey, 2002, p. 52).

Conscious, Participatory Agents

"Even if the human mind, along with the individual personality associated with it is illusory, it is this illusion with all its intricacies,

---

90 In these pages, we are talking of consciousness in a somewhat general way—from the consciousness in quantum particles, to the consciousness of a rock, to that of plants and animals and humans, all on a developmental continuum. That continuum persists in humanity, where we have identified developmental levels (which we'll address later). We are also including self-reflective consciousness, the capacity to know that we know, in this general conversation. We'll also have much more to say about that, in particular, as we go.

quirks, and faults that decides all issues," says Wallace. "It is the mind that becomes foolish or wise" (Wallace, 2008, p. 118).

It seems, then, as self-reflective conscious beings, mindful beings, we have become perceptive participants in shaping what becomes of the story of the universe. We have become co-creators—knowers of what exists, capable of reflecting on it all and of knowing that we know. We are agents of movement and action, observers and witnesses of our world, interpreters of meaning and purpose, doers and designers of our destiny, pursuers of our preferences, and self-determiners acting on our values. We are movements within the greater movement of the cosmos.

Kauffman remarks in amazement: "It is utterly remarkable that agency [that is, choice based action] has arisen in the universe—systems are able to act on their own behalf; systems that modify the universe on their own behalf. Out of agency comes value and meaning" (Kauffman, 2007, pp. 903-914).

Physicist Amit Gotswami sees that this "meaning arises in the universe when sentient beings observe it, choosing causal pathways from among the myriad transcendent possibilities" (Gotswami, 1993, p. 141). Meaning arises not from matter-based insentience, but rather, from the "myriad and transcendent possibilities" of consciousness.

Kauffman revels in this mystery of consciousness, meaning, and choice, manifesting in our universe: "Life is valuable on its own, a wonder of emergence, evolution and creativity. Reality is truly stunning" (Kauffman, 2007, pp. 903-914).

What happens in the world is predicated on *observation,* and because of our capacities for self-reflective consciousness and agency, we affect how the universe unfolds. This means that the universe is *participatory* (as the celebrated physicist Wheeler says), and it acts in ways which are not strictly material cause and effect. "Humans and particles—and all other persons—are conscious, participatory, mindful, material and organisational," affirms Harvey in his insightful book, *Animism: Respecting the Living World* (Harvey, 2005, p. 202).[91]

---

91 The philosopher Henri Bergson, observed, "Intuition is mind itself, and, in a certain sense, life itself." (Bergson, 1998, p. 268), and poetically as well as philosophically, amplifies on this notion of agency, cautioning that we not privilege intellect over other types of knowing, other ways of expressing free agency: "The movement of the stream is distinct from the riverbed, although it must adopt its winding course. Consciousness is distinct from the organism that it animates, although it must undergo its vicissitudes…

Coupled with Einstein's relativity theory—which, again, says that time and space are not fixed realities, but unique to each person's experience—we can truly say that we live in a *participatory* universe which acts in ways that are often more of an *organic process* than mechanistic cause and effect, and within which *each individual's perception* is uniquely shaped.

This notion of a solely mechanically driven universe is an antiquated one: it deprives us of appreciating the great depths of perceiving and knowing, reflection, awareness, intelligence, relationship, cooperation, compassion, and creativity we have become capable of being, and it cheats us of a regard for the impressive deepening we might arrive at as a species. It denies us of entertaining a greater vision of the potential that these attributes offer us as we grow forward.

In an interview by Alan Steinfeld for *New Realities*, Chopra makes the following observations about the phenomenon we call consciousness: "Right now we're perceiving each other, talking to each other. And we seem to think that we are doing it. But it's actually an activity of the universe...It requires all the dark energy, all the dark matter, and the rotation of the earth on its axis, and moving around the Sun, as it does, to have this moment. So *consciousness is an activity, this second, of the total universe* [emphasis added] ...If the total universe didn't exist as it exists, there would be no awareness...."

The fundamental essence of the universe itself, he says, is consciousness: "There is nothing 'out there.' There is only 'in here.' And the 'in here' experiences itself as the 'out there' as well...There is in fact no such thing as a person. There is only the universe pretending to be a person... There is only one 'I'" (Steinfeld, Alan, 2012, NewRealities.com).

So, "Because consciousness transcends the body, because *internal* and *external* are fundamentally distinctions of language and practicality alone," says Lanza, "we're left with Being, or consciousness as the bedrock component of existence" (Lanza, 2009, p. 189).

"The destiny of consciousness is not bound up... with the destiny of cerebral matter.... Consciousness is essentially free; it is freedom itself; but it cannot pass through matter without settling on it, without adapting itself back toward active, that is to say free, consciousness, naturally makes it enter into the conceptual forms into which it is accustomed to see matter fit... It will therefore always perceive freedom in the form of necessity; it will always neglect the part of novelty or of creation inherent in the free act; it will always substitute for action itself an imitation artificial, approximative, obtained by compounding the old with the old and the same with the same" (Bergson, 1998, p. 270).

Further, Lanza continues, "Some may imagine that there are two worlds, one 'out there' and a separate one being carbonized inside the skull. But the 'two worlds' model is a myth. Nothing is perceived except perceptions themselves, and nothing exists outside of consciousness. Only one visual reality is extant, and there it is. Right there.

"The 'outside world' is, therefore, located within the brain or mind" (Lanza 2009, p. 36).[92]

Lanza,[93] thinks that the entire cosmos is a figment of our imagination, that is, it exists only in our minds as an active process that actually involves our consciousness, a theory he calls biocentrism.

His conclusion is that "It's actually us, the observer who creates space and time, and that's why you're here, now. Reality begins and ends with the observer" (Lanza, 2009, Jacket).

Conscious observation, mind, it seems, does create our reality!

REALITY IS A LIVING, SUBJECTIVE EXPERIENCE

"The philosophy of science that ignored the question of consciousness is standing there dead. Science can no longer ignore the fact that our conscious observation affects the quantum potentialities of matter, and we can no longer ignore the significance of what our will does in bridging minds and affecting phenomena in the physical world."

—Evan Harris Walker (Walker, 2000, p. 331).

We live in a subjective universe. "No human—scientist or

---

92 Lanza observes, "Our entire education system in all disciplines, the construction of our language, and our socially accepted 'givens'—those starting points in conversations—revolve around a bottom-line mindset that assumes a separate universe 'out there' into which we each individually arrived on a very temporary basis. It is further assumed that we accurately perceive this external pre-existing reality and play little or no role in its appearance" (Lanza, 2009, p. 15).

93 According to the Huffington Post, Dr. Robert Lanza is considered one of the leading scientists in the world. He is currently Chief Scientific Officer at Advanced Cell Technology, and a professor at Wake Forest University School of Medicine. He has several hundred publications and inventions, and over two dozen scientific books: among them, *Principles of Tissue Engineering, which* is recognized as the definitive reference in the field. Others include *One World: The Health & Survival of the Human Species in the 21st* Century (Foreword by President Jimmy Carter), and the *Handbook of Stem Cells and Essentials of Stem Cell Biology,* which are considered the definitive references in stem cell research. Dr. Lanza received his BA and MD degrees from the University of Pennsylvania, where he was both a University Scholar and Benjamin Franklin Scholar. What he is saying here, therefore, should not be trivialized or dismissed: "Everything you see and experience is a world of information occurring in your mind."

other—has ever had any experience outside of consciousness or outside of the eternally present moment," notes evolutionary biologist Sahtouris, (Sahtouris, 2000, p. 376), pointing us to the unarguable conclusion that *it is not an objective viewpoint that is the cornerstone of reality, but a subjective one*, because this eternally present moment is experienced uniquely, subjectively, and ultimately, privately, by each individual.

However, for space-time to be subjective, there had to be some form of awareness or consciousness or sense of self-presence existing from before the moment of the big bang, the singular event that resulted in the dimensions we call time and space—or space-time—would not exist, (i.e., it would be unknown). If subjective awareness—again, this is the conclusion of quantum theory—is what underlies our universe, then a conscious (living) process must, somehow, have been going on throughout.

Sahtouris argues, "Consider a universe of pure energy with the appearance of material reality. To have an *appearance* there must be an observer, and as quantum theorists pointed out long ago, in a completely interconnected universe, consciousness anywhere means consciousness everywhere. Now non-locality tells us that anywhere is everywhere! In fact it would seem that energy itself, like matter, is an 'appearance' of consciousness" (Sahtouris, 2000, p. 376).

Indeed, some form of consciousness, some subjective presence, must have existed from "the beginning of time." So we see that the notion of a "dead," inert universe inexplicably filled with lonely matter that mysteriously results in chemical interactions called "life" and "mind" must be put to bed.

Challenging the deeply held but unproven and even magical tenet of scientific materialism—that matter is primary—de Quincey contends that "mind cannot emerge from wholly insentient matter... it is inconceivable that subjectivity could emerge from pure objectivity... The universe would always and forever be unexperienced and unknown" (de Quincey, 2005, pp. 264-265).

What it is that is subjectively experienced, subjectively known, is *information*. In fact, it is information that is the stuff of consciousness. Wallace affirms, "It is *information* that determines the physical system.

The very characteristics that make data "information" are determined by the human mind. It is we who determine which data are relevant information, and which are only background noise. This alone suggests that the mind is an intimate participant in physical events. It is not closed off to them" (Wallace, 2008, p. 50).

"Physical objects are not fundamental," says Hoffman. "Space-time is doomed. There is no such thing as space-time, fundamentally, in the actual underlying description of the laws of physics. That's very startling because what physics is supposed to be about is describing things as they happen in space and time.

"So 'if there's no space-time, it's not clear what physics is about,' says Nima Arkani-Hamed, in the Cornell Messenger Lecture 2016. And this is one of the best and brightest in the field," says Hoffman. "And what we know is that for us to get a theory of quantum gravity, we have to let go of the notion that space-time is fundamental" (Hoffman, 2017, *Simulation*).

Again, then, we find that physical experiences are illusory, products of the information determined by subjective minds. "When talking about the mind or consciousness, we should use words and ideas that refer to connections through *meaning*, not connections through mechanism," affirms de Quincey. "Energy talk about mind fails to account for the most fundamental characteristic of consciousness: its *subjectivity*. The feeling of consciousness has no objective, measurable, position in space. It is not even 'nonlocal'… It is, *rather, non-located*—it is not located in space in any way whatsoever, either as 'fields,' 'vibrations,' or 'waves,'" (de Quincey, 2005, p. 54)—because "There is no such thing as space-time."

Other thinkers, de Quincey points out, have, as well, established that "the raw 'stuff' of the universe, its primordial 'matter,' must always have possessed an 'interiority' (Wilber), a 'within' (de Chardin), an 'occasion of experience' (Whitehead), a "quantum of consciousness' (Young), 'a capacity for a subjective viewpoint' (Nagel), 'a self-organizing intelligence' (Bruno). Matter, in other words…is, and was always, intrinsically intentional, significant, and meaningful" (de Quincey 2005, pp. 264-265). What we call "matter," indeed, experiences.

"You come to a view of a living universe," Sahtouris says, "rather than this strange concept among human cultures that western science came to that we're a non-living universe—a mechanical, celestial-me-chanical if you like—that's running down by entropy,[94] and in which, by some miracle, life emerged from non-life, consciousness from non-consciousness, intelligence from non-intelligence—and those have been the stickiest problems for western science" (Sahtouris, *After Darwin, Part 1*).Quantum theory holds that observation—subjective observation—must occur for us to experience existence. Observation means consciousness. Consciousness is therefore required for reality—all that we experience—to exist.

So the cosmos appears to be conscious as a subjective presence, in a dynamic process, unfolding as an undivided movement we perceive as space-time and demonstrating ever more complexity and self-reflective capacity.

Walker points out, "Consciousness is not so many atoms. It does not consist of photons or quarks. Neither is it molecules spinning about in the brain. Consciousness is something that exists in its own right" (Walker, 2000, p. 182).

Nick Herbert envisions "a kind of 'quantum animism' in which 'mind permeates the world at every level:' with consciousness, a 'fundamental force enters into necessary cooperation with matter to bring about the fine details of our everyday world,'" he says (Herbert, 1993, p. 187).

Chopra's perspective is this: "You and I are part of a conscious universe. The universe is thinking. It's creative. It imagines. The universe is full of creativity, and it couldn't be full of creativity if it wasn't conscious. I am creative because the universe is creative. I am conscious because the universe is conscious. I think because the universe thinks. I am imbued with subjectivity because the universe is imbued with subjectivity, which means the universe has a sense of 'I am. I exist.' My sense of 'I am,' my sense of my existence, is not separate from the universal sense of 'I am'" (Chopra, 2006, p. 130).

---

94 A term taken from thermodynamics that refers to the process of the universe to generally move toward greater degradation, running down, or tending toward increasing disorder.

How Are "Matter" and "Consciousness" Interconnected?

"Our current theories of the physical world don't work, and can never be made to work until they account for life and consciousness."

—Robert Lanza, (Lanza, 2010, p. 8)

The existence of consciousness begs one of the most intriguing questions "of all time:" How are "matter" and "consciousness" interconnected?

Philosopher de Quincey, in tackling this question, poses four options, and comes to a conclusion for himself about each:

1. *Is only matter real (and consciousness merely a manifesting of parts and processes contained in the brain)?*

This is the "materialist" position that science has assumed for three hundred and fifty years. "If only matter and physical energy is ultimately real…the problem is how to explain 'something' that has no mass, occupies no space, and has subjectivity could ever evolve or emerge from something that was massive, spatial, and wholly objective to begin with" (de Quincey, 2002, p. 267).

"For materialism, the difficulty is in accounting for the emergence of sentience and subjectivity wholly from insentient and objective matter" (de Quincey, 2002, p. 182).

2. *Is the existence of consciousness completely separate and independent from matter?*

This is "Cartesian dualism." "Both matter and mind are real, but they are different substances and exist separately…If this were the case, we would be left with the unyielding problem of explaining how two mutually alien substances could ever interact," says de Quincey (de Quincey, 2002, p. 42). However, there has to be an interface between matter and consciousness, because we experience consciousness influencing matter all the time. Walker proposes such a "mechanism" in his book, *The Physics of Consciousness:* there is indeed such an interface. It occurs at a synapse in the brain where subatomic quantum particles interface with the physical brain (Walker, 2000, pp. 194 ff.). So perhaps Descartes could not be too easily dismissed if what we perceive as matter truly existed as such.

3. Is nothing but Spirit, God, mind, or consciousness real?

In other words, is consciousness primary? This is "idealism," also known as the perennial philosophy. In idealism, only mind, consciousness, or spirit is ultimately real.

In this perspective, "Consciousness is primary and universal, and… matter is either (a) merely an illusion, or (b) an emanation of spirit," says de Quincey. "According to absolute idealism, the nature of ultimate reality is spirit or pure consciousness, and the world of matter is an illusion, a sort of cosmic dream, what the Hindu tradition calls *maya…*

"Another version, emanationist idealism, is based on…[the] premise that the cosmos is structured according to a hierarchy of ontological levels, from matter, body, mind, soul, to spirit" (de Quincey, 2002, p. 44).

"The problem with idealism," claims de Quincey, "is logically less severe than with either dualism or materialism, but nonetheless, needs to be addressed: If all is spirit, and matter is ultimately illusion or manifestation of spirit, how do we account for the universal, commonsense, and pragmatic supposition of realism—that the world is real in its own right?" (de Quincey, 2002, p. 44). If our experience is that the world is real, we naturally want to operate out of that mindset.

When I consider this point of view, I find that de Quincey dismisses the merits of idealism a bit out of hand. Quantum theory has concluded that matter *is* illusory, and we (and he) have made the point that the underlying reality is quantum dynamics, indeed, consciousness, which generates, or creates, matter. Science says that in the infinitesimal vacuum at the time of the big bang were "quantum fluctuations," and quantum theory holds that, as matter is illusory, consciousness must be first, *and* that observation is requisite for the material world to manifest. That results in a strong correlation between consciousness and quantum fluctuations. Indeed, we will see that they are equivalent, what we can call the quantum mind.

Additionally, Lanza, as we've seen, "firmly believes the universe begins and ends in the mind of the observer, that the universe is nothing more than the vivid imagination of our brains." Of course, we can't prove otherwise, so pick your poison. But he argues as we've noted, that, "…Our current theories of the physical world don't work, and can never be made to work until they account for life and consciousness.… Rather than a belated and minor outcome after billions of years of lifeless physical processes, life and consciousness are absolutely fundamental to our understanding of the universe" says Lanza (Lanza, 2009, p. 2).

He points out persuasively, "Consciousness is not just an issue for biologists; it is a problem for physics. Nothing in modern physics explains how a group of molecules in your brain create consciousness… Nothing in science can explain how consciousness arose from matter. Our current model simply does not allow for consciousness, and our understanding of this most basic phenomenon of our existence is virtually nil. Interestingly, our present model of physics does not even recognize this as a problem" (Lanza, 2009, p. 4).

"'Theories of Everything' that do not account for life or consciousness will certainly lead ultimately to dead-ends, and this includes string theory," Lanza further contends. "Models that are strictly time-based, such as further work on understanding the Big Bang as the putative natal event of the cosmos, will never deliver full satisfaction or closure" (Lanza, 2009, p. 162).

"In reality, when we are scientifically honest," he declares, "we must admit we have no idea how awareness can ever arise—not in an individual, not collectively, and certainly not from molecules and electromagnetism. Indeed, does consciousness arise at all? It's widely repeated that each cell in our body is part of the continuous string of cells that started dividing billions of years ago—a single unbroken chain of life. But what about consciousness? This more than anything else must be unbroken. Although most people like to imagine a universe existing without it, we have seen that this makes no sense if one gives the matter

A Universe Full of Magical Things

sufficient thought. How does consciousness ever begin? How could that possibly occur?" (Lanza, 2009, p. 178).

Lanza quotes physicist Andrei Linde of Stanford University, "the universe and the observer exist as a pair. I cannot imagine a consistent theory of the universe that ignores consciousness...." (Lanza, 2009, p. 178).

It seems that the idealist point of view may, after all, be the one that best reflects "reality," especially a non-physical, illusory, virtual reality, which must be the case if there is no such thing as matter per se, but instead only action, as quantum theory holds.

Recall Physicist Dyson's perspective: "Matter in quantum mechanics is not an inert substance but an active agent, constantly making choices between alternative possibilities...It appears that mind, as manifested by the capacity to make choices, is to some extent *inherent in every electron*" [emphasis added] (Frankenberry, 2008, p. 372). It seems that mind, i.e. consciousness, is what is real, what is "the thing."

It isn't that consciousness emanates from matter; rather consciousness passes through matter.

Elgin elaborates: "This does not mean that an atom has the same consciousness as a human being, but rather that an atom has a sentient capacity appropriate to its form and function" (Elgin, "Our living," December 2011). Yet, mind pervades matter down to the quantum particle, and matter is illusory. The answer to de Quincey's question, "If all is spirit, and matter is ultimately illusion or manifestation of spirit, how do we account for...the supposition of realism?" seems to be contained in the question itself: All *is* mind—spirit—and matter is ultimately illusion or manifestation of spirit.

4. *Or, have life and consciousness and subjectivity in some form co-existed and co-emerged with matter from the beginning?*

This is the fourth option de Quincey proposes be considered, "radical naturalism," or panpsychism, the solution he advocates. He says, "It is inconceivable that sentience (subjectivity, consciousness) could ever emerge or evolve from wholly insentient (objective, physical) matter."

He continues, "... (If) consciousness exists now, it must always have existed in some form...Similarly, if matter exists it must always have existed in some form. The central tenet...is that matter is intrinsically sentient—it is both subjective and objective...or consciousness within matter" (de Quincey, 2002, pp. 45-46). "We must accept that unless consciousness is something which just suddenly emerges, just gets added on with no apparent cause, then it was there in some form all along as a basic property of the constituents of matter," comments management thought leader, physicist, philosopher, and author Danah Zohar, aligning with de Quincey (Harvey, 2005, p. 189).

In this view, matter and consciousness are co-emergent; subjectivity and objectivity are intrinsic to reality. Neither emerged out of the other, they co-emerged, contemporaneously.

But how can that be, since quantum theory states that there is no objectivity: only subjective perceptions of objectivity exist in individual, unique experience? Objectivity doesn't exist, except as an idea in our minds, as a convention in our communications, and as a cultural agreement of convenience. The 'objective' procedures and the structures upon which our subjective perceptions operate are not a thing.

Either matter exists as a real thing or it exists only as "packets of action," quanta, and has no physicality. Quantum theory says the latter. So, of course, consciousness exists in matter and beyond, but matter only exists, subjectively, in consciousness.

We can *say* that both matter and consciousness have co-existed from the beginning. *In our experience,* each requires the other. *For us,* matter has no meaning without co-evolving consciousness, and consciousness has no meaning without co-evolving matter. So panpsychism or natural realism suffices, FAPP (For All Practical Purposes). Yet the overarching reality is the non-physicality of quantum. The underlying quantum structure of the cosmos is consciousness, (quantum/mind), and it is not physical.

This means that the universe has an innate aliveness, subjective awareness, and intentionality. It means the universe, and our being here, are not random, not meaningless. It means we have a purpose that was present for us from before the "time" when we came into temporal existence. And now, it means that we may be growing closer to understanding exactly what that purpose could be.

Chopra from his own perspective adds this insight: "To separate the body-mind from the rest of the cosmos is to misperceive things... The body-mind is part of the larger mind; it's part of the cosmos" (Chopra, 2006, p. 43). The answer to our quest for purpose resides there.

Chopra tells us further: "Everything in the universe is alive. The Earth, the stars, Milky Way, and other galactic systems are a living organism. The universe is one gigantic living Being. When we have a feeling of intimacy with this Being, when we fall in love with everything that exists, the universe speaks to us and reveals its innermost secrets" (Chopra, 2006, p. 129). In this statement, we can take a hint about the far deeper meaning and purpose of our universe than materialism might ever imagine!

Consider once more Sahtouris' perspective: "Consider a universe of pure energy with the appearance of material reality. To have an appearance, there must be an observer, and as quantum theorists pointed out long ago, in a completely interconnected universe, consciousness anywhere means consciousness everywhere. Now non-locality tells us that anywhere is everywhere! In fact, it would seem that energy itself, like matter, is an 'appearance' of consciousness... Thinking things through in this way, we see how limited our worldviews have been" (Sahtouris, 2000, p. 376).

Blogging on the kwelos.tripod.com website, Sean Robsville argues: "We can dismiss all claims that consciousness, mind and awareness are emergent properties of matter or brains, because we need the presence of a mind for emergent properties and phenomena to appear in the first place. The subjective activity of the mind of the observer, together with the 'objective' procedures and the structures upon which they operate, is an irreducible component of emergent phenomena" (Robsville, kwelos.tripod.com/emergent.htm). Mind before matter.

So we return to idealism, the point of view that consciousness is primary—but with a twist, which de Quincey does not address: how might idealism look in our world? That is the subject of the next chapter, where we'll look at "conscious realism," as described by Hoffman, and at the closely related concept of "quantum realism," a notion promoted by Brian Whitworth.

Quantum theory has shown that we humans are participants in

shaping what is created: we are co-creators with the universe. We know the universe is not strictly cause and effect: there is a non-mechanical element, a vital, emergent, active process of unfolding, in which we play a conscious and subjective part. We are participatory agents, co-creators—*what we grow to be conscious of matters, and what we choose to be conscious of matters, because it shapes what the universe, and we, become!*

AN EXPLOSIVE STATE CHANGE

Because of this new paradigm we are talking about, it will be possible to look at ourselves, our world, and our purpose much more deeply. Sahtouris states "…there will be enormous effects of learning that our consciousness creates our reality—that our assumptions, our beliefs as individuals, as society, and as humanity are the basis for the world we produce for ourselves and co-produce together, along with all living systems, from moment to moment" (Sahtouris, *Darwin Part 1*).

We exist in relationship with the cosmos because of our conscious awareness. It is a new threshold, a new game entirely: Our "'genes, in making possible the development of human consciousness,'" says Richard Lewontin, "'have surrendered their power both to determine the individual and its environment. They have been replaced by an entirely new level of causation, that of social interaction with its own laws and its own nature'" (Gould, 2009, p. 120).

"The biological evolution" pronounces Theodosius Dobzhansky, "has transcended itself in the human revolution" (Dobzhansky, 1967, p. 58).

According to Lipton, the new science of epigenetics, (i.e. "control above genetics,") "profoundly changes our understanding of how life is controlled…Epigenetic research has established that DNA blueprints passed down through the genes are not set in concrete at birth. Genes are not destiny! Environmental influences, including nutrition, stress, and emotions, can modify those genes without changing their basic blueprint. And those modifications, epigeneticists have discovered, can be passed on to future generations…." (Lipton, 2005, p. 37).

We are finding that, since the development of a depth of

consciousness that involves intelligence, self-reflection, compassion, and creativity—among other distinctive features—no longer do our genes alone determine the course of our future.

Our ability to imagine, plan, to think in abstract ways, to feel, and to communicate from human to human; all congeal to impel us, in harmony with, and yet beyond, the potential of our genetic makeup. We now move to an unknown world where our creative potential both transcends and includes our DNA, that adaptive and intelligent agent of our evolution that has brought us to here.

Rolston notes about the emergence of mind in humans: "So the puzzle is how a change of some one or two percent in DNA results in light-years of mental explosion...Perhaps the radical threshold is crossed with the emergence of consciousness.... Although consciousness preceded humans, there was an explosive state change when humans crossed the divide and gained their self-reflective, ideational, linguistic, symbolic capacities. Humans are not simply conscious; they are self-conscious [i.e. self-reflective] ... The key threshold is the [intersubjective] capacity to pass ideas from mind to mind," (Rolston, 2010, p. 89), that is, the capacity to have a conscious, creative relationship with another, and with others.

As we've seen and will see, *we exist in a living universe (and on a living Earth). The cosmos is an alive, intelligent, conscious and relational unfolding. We are involved in an emerging, evolving, intentionally creative process of being and becoming.*

We've seen the advances in physics and chemistry resulting from the insights of the theories of quantum and relativity. We can see from biology and epigenetics the transformative understanding we now possess of living processes and the interdependency that defines life itself—which we will examine more thoroughly in Chapters Seven, Eight, and Nine. We've seen that *the universe is a vital creative process*, not an inert, inanimate machine. Indeed, the truths of the cosmos are possibly more like what we would consider science fiction from a culturally orthodox worldview. "Quantum physics tells us that the world is composed of an underlying field of intelligence," says Chopra, "that manifests as the infinite diversity of the universe" (Chopra, 2006, p. 26).

The Mind and the Brain

"Physics, the most mature of all the natural sciences, has been forced by experimental evidence to include the role of the observer as a complementary aspect of physical reality. Mind and matter seem to form an inseparable pair, in essence, an interconnected universe. Meanwhile, the other sciences, supposedly basing themselves on the conclusions of physics, still equate mind with the brain—a purely physical reality, if even that. Cognitive scientists in particular remain entranced by the illusions of scientific materialism, some no doubt, dreaming of a day when a sophisticated computer will actually 'become human,' or at least conscious."

—B. Alan Wallace (Wallace, 2008, p. 109).

The relationship between the mind and the brain is, of course, understood based on one's particular (subjective) theory about the essential nature of being or existence, one's ontological viewpoint. Physicalists believe consciousness has direct, comprehensive, and specific correlations in the physical brain, that mind (consciousness) is a function of the material brain. Dualists hold that the non-physical mind and physical brain are separate, yet somehow interface with each other. Natural realists (panpsychists) believe mind and brain co-evolved, that consciousness and matter co-evolved. And idealists hold that all is consciousness and that the physical world—and the brain—are illusions of the mind.

One version of idealism, which refutes the belief that ultimate reality is what we perceive, is called quantum realism, (or conscious realism). Quantum realism considers reality to be virtual—in effect, the result of a spectacularly inventive programmer and a cosmos that was "booted up" at the big bang (or at the inception of a metaverse).

We've seen that data from quantum dynamics and the resulting metaphysical considerations we've explored, lead us, despite what reductionist science affirms, to an ontological conclusion that consciousness *is* the ultimate nature of the cosmos, with all that we experience being artifacts of this consciousness, the quantum mind.

Despite quantum insights that negate the primacy of matter,

neuroscience continues to operate on the assumption that consciousness arises out of a material substance, the brain. Yet, according to Wallace, "By assuming that mind equals brain—that all subjective states boil down to physical functioning—brain science has placed itself in a difficult position. If mental phenomena are emergent or secondary processes of brain tissue, their relationship to the brain is fundamentally different from every other emergent property and function observed in nature... In the theory of brain correlates to mental states...observation is missing: no mental states are observed when examining brain tissue, and conversely, the brain is not observed as part of the experience of a mental state. The idea that mental processes and their correlated brain functions are equivalent is a hypothesis that is simply not validated, yet it is one that scientific materialists find appealing and therefore often accept without question" (Wallace, 2008, p. 81). Neuroscience clings tightly to a materialistic correlation of thought and brain tissue, a bias that has not been verified, and so is vulnerable to false assumptions and conclusions.

Just as if you search all day inside a radio you will never find Dion's *Runaround Sue* or Handel's *Messiah*, so too with consciousness and the brain. You can search all day in the brain and you will never find consciousness. *Runaround Sue* passes through the radio, and consciousness passes through the brain. Indeed, it isn't that all of consciousness emanates from matter; rather consciousness passes through all of matter.

Eric Kandel, one of the giants of "the new science of mind," and a Nobel Prize winner, in his influential book, *In Search of Memory*, aptly and liberally articulates the materialistic point of view that thought and brain tissue are correlative, stating that "The new science of mind attempts to penetrate the mystery of consciousness, including the ultimate mystery: how each person's brain creates the consciousness of a unique self and the sense of free will" (Kandel, 2006, p. 11). Notice Kandel makes the *assumption* that "each person's brain creates the consciousness of a unique self and the sense of free will."

He describes the evolution of this discipline thus: "In the 1970s, cognitive psychology, the science of mind, merged with neuroscience, the science of the brain. The result was cognitive neuroscience, a discipline that introduced biological methods of exploring mental processes into modern cognitive psychology. In the 1980s, cognitive neuroscience received an enormous boost from brain imaging, a technology that enabled brain scientists to realize their dream of peering inside the human brain and watching the activity in various regions as people engage in higher mental functions…Brain imaging works by measuring indices of neural activity: positron-emission tomography (PET) measures the brain's consumption of energy, and functional magnetic resonance imaging (fMRI) measures its use of oxygen. In the early 1980s cognitive neuroscience incorporated molecular biology, resulting in a new science of mind—a molecular biology of cognition—that has allowed us to explore on the molecular level such mental processes as how we think, feel, learn, and remember" (Kandel, 2006, pp. 7-8).

These tools, which measure physical attributes of the brain, powerfully illuminate how the physical brain functions. But neuroscientists have still not observed one single mental state, and yet they regard their findings to be reflective of mental states: they regard mental states as physical entities because they're able to observe physical dynamics of the brain. They operate on an unproven assumption, a belief, not a verified fact! It's as if we watch something on TV and because we see movement and story, we conclude that we are observing something real.

Kandel continues, citing Darwin's *belief*: "Charles Darwin argued that we are not uniquely created, but rather evolved from lower animal ancestors…He proposed the even more daring idea that evolution's driving force is not a conscious, intelligent or divine purpose, but a "blind" process of natural selection, a completely mechanistic sorting process of random trial and error based on hereditary variations…

"Eventually, modern biology would ask us to believe that human beings, in all their beauty and infinite variety, are ever new combinations of nucleotide bases, the building blocks of DNA's genetic code" says Kandel (Kandel, 2006, p. 8).

Then, he concludes his description of the roots of neuroscience with another *belief*: "The new biology of mind is potentially more disturbing because it suggests that not only the body, but also the mind and the specific molecules that underlie our highest mental processes—consciousness, of self and of others, consciousness of the past and the future—have evolved from our animal ancestors. Furthermore, the new biology posits that consciousness is a biological process that will eventually be explained in terms of molecular signaling pathways used by interacting populations of nerve cells" (Kandel. 2006, pp. 8-9).

The philosophy of the new science of mind embraces fully and without examination the tenets of reductionist and mechanistic science, as Kandel's words sharply document. He argues that "the idea that the human mind and spirituality originate in the physical organ, the brain, is new and startling for some people. They find it hard to believe that the brain is an information-processing computational organ made marvelously powerful not by its mystery, but by its complexity—by the enormous number, variety, and interactions of its nerve cells" (Kandel, 2006, p. 9). This, be clear, is a belief system, and its tenets are completely speculative, undocumented, and sadly, unthinkingly aligned with mechanistic verses quantum principles—aligned with the Materialistic Paradigm of matter first.

Certainly, neuroscience has made surprising inroads: "As theories, the work of the past quarter-century reflects some of the important progress that is occurring in the field of neuroscience and psychology," Lanza acknowledges. "The bad news is that they are solely theories of structure and function. They tell us nothing about how the performance of these functions is accomplished by a conscious experience. And yet the difficulty in understanding consciousness lies precisely here, in this gap, in understanding how a subjective experience emerges from a physical process at all. Even the Nobel Laureate physicist [and outspoken atheist] Steven Weinberg concedes that there is a problem with consciousness, and that although it may have a neural correlate, its existence does not seem to be derivable from physical laws" (Lanza, 2009, pp. 173-174).

Thus, although the science of neurobiology *has* made great strides in understanding conscious experience with respect to some of its correlated *structural and functional physical expressions* in the brain, materially-oriented neurobiologists unfortunately make the leap from correlation to causation: they see a physical manifestations in the brain, which, they hold, "causes" our conscious experience.

Kandel references two "characteristics [of] the conscious state, unity and subjectivity. [The first is] the unitary nature of consciousness [which] refers to the fact that our experiences come to us as a unified whole…

"Subjectivity, the second characteristic of conscious awareness," he continues, "poses the more formidable scientific challenge. Each of us experiences a world of private and unique sensations that is much more real to us than the experiences of others…The fact that conscious experience is unique to each person raises the question of whether it is possible to determine objectively any characteristics of consciousness that are common to everyone. If the senses ultimately produce experiences that are completely and personally subjective, we cannot…arrive at a general definition of consciousness based on personal experience" (Kandel, 2006, pp. 279-280).

Coming from his materialistic, objective bias, Kandel says, "What science lacks are rules for explaining how subjective properties (consciousness) arise from the properties of objects (interconnected nerve cells)" (Kandel, 2006, p. 281). However, according to quantum theory, neither objective properties nor physical objects exist, so the answer, assuming a Materialistic Paradigm, will be difficult to arrive at. Alternately, if one assumes a Consciousness Paradigm, the answer is easy.

Eben Alexander was a practicing neurosurgeon until he collapsed on his bed in a coma in November, 2008. "According to conventional science, due to the severe damage in my brain caused by an overwhelming bacterial meningoencephalitis, I should not have experienced anything—at all," he said (Alexander, 2017, p. xiii).

He continued, "By day three of such an illness, virtually all patients are beginning to awaken, or they're dead." Hope for a positive outcome "dwindled to a 2 percent chance of survival after a week spent in a coma. Much worse…was the harsh reality [the treatment team] attached to it, and that was the likelihood of my actually awakening and having some return of quality to my life. Their estimate for that possibility was a most disappointing zero—no chance of recovery to any sort of normal daily routine" (Alexander, 2017, p. 3) … "I had discovered no cases of any other patients with my particular diagnosis that went on to benefit from complete recovery" (Alexander, 2017, p. 4).

While in a comatose state, Alexander reported rich and vibrant mental experiences he describes in his two books, *Proof of Heaven,* and *Living in a Mindful Universe.* "Those who had taken care of me, and those who knew enough neuroscience to understand the impossibility that such an impaired brain could have even remotely offered up that extraordinary and detailed complexity of experience, shared a much deeper sense of mystery," followed by bewilderment at his subsequent complete recovery, which was extensively and broadly documented.

"Over three decades of my life spent working daily with neurosurgical patients...I had come to believe that I had some understanding of the relationship between the brain and mind—the nature of consciousness" Alexander says. "Modern neuroscience has come to believe that all of our human qualities of language, reason, thought, auditory and visual perceptions, emotional forces, etc.—essentially all the qualities of mental experience that become part of human awareness—are directly derived from...the neocortex..."

However, he acknowledged, "I accepted the neuroscience party line that the physical brain creates consciousness out of physical matter" (Alexander, 2017, p. 5).

"The general approach to a case like mine had been to sweep it under the rug, out of the way, and simply accept it as unexplainable. But my confidants understood my dilemma and supported my quest to more properly comprehend it. There was something much greater going on here, and I was driven to seek deeper understanding" (Alexander, 2017, p. 12).

He came to the conclusion that "Some of the modern scientific thinking now sweeping the field of consciousness studies concerns a wholly different concept of the mind-brain relationship: that the brain is a reducing valve, or filter, that reduces *primordial* consciousness down to a trickle—our very limited human awareness of 'here and now.' The physical brain only permits certain patterns of awareness to emerge from a broad group of possible mental states. This conscious awareness can be liberated to a much higher level when freed up from the mental shackles of the physical brain, as happened when I was in a coma" (Alexander, 2017, pp. 22-23). Regardless of his new perspective, it was clear: zero brain functioning in medical patients *never* results in complete recovery...yet, in Alexander's case, it had! The data from his extremely thoroughly verified experience unequivocally demonstrate that personal consciousness cannot be located in the physical brain.

"The inability to identify any physical location of memory in the brain is one of the greatest clues that materialism is a failed worldview. The more we learn about the structure and biology of the brain, the clearer it becomes that the brain does *not* create consciousness, nor serve as the repository for memory," says Alexander. "The brain doesn't produce consciousness any more that it produces sound waves when you hear music. In fact, the situation is just the opposite: We are conscious *in spite of* our brain" (Alexander, 2017, p. 49).

Hoffman unequivocally concludes, "When we've tried to solve the problem of consciousness and the mind-body problem by assuming that brain activity causes our conscious experiences, we've gotten nowhere. No one has been able to figure out how brain activity could cause our conscious experiences. *There's actually nothing intelligent to say along that line* [emphasis added]. And the reason why is because space and time are just fictions that we use to stay alive and physical objects are useful representations, but have no causal powers.

"So neurons literally have no causal powers. Your brain causes none of your perceptions, none of your behaviors, and none of your emotions. Brain activity causes nothing. There is something that is causing all of that stuff...Some other theory needs to be put out there because the space and time of physical objects have no causal powers. They cannot explain it...

"When we're actually trying to understand consciousness and the mind/body problem, we have to look at our assumptions...All of the theories that you've heard about from most people...they all assume the reality of physical objects in space and time. And they assume they have genuine causal powers...And what we are saying here is, absolutely not. There is no causality in space and time... so there's nothing... to grab hold of. There are no causal structures'" Hoffman insists (Hoffman, 2017, *Simulation*). We'll hear more about this line of thought from Hoffman in the next chapter.

Wallace likewise maintains, "Brain science has no technological means of detecting consciousness or any kind of subjective mental

A Universe Full of Magical Things

experience. Nor does it have any means of demonstrating how the brain produces consciousness. By relying on the argument of mere correlations between mental phenomena and brain physiology, cognitive psychologists remind us of astrologers, who rely on the correlates between patterns in the heavens and events on Earth, rather than astronomers, who have actually explored the sky scientifically with telescopes" (Wallace, 2008, p. 82). A Materialistic Paradigm cannot address that which it does not acknowledge and cannot see.

Steve McIntosh pursues the point—that the science of neuroscience is off base by assuming a physicalist hypothesis—further: "Despite the recent advances in neuroscience, when it comes to the study of consciousness, the scientific materialists have a crucial problem—it's called the 'mind/body problem,' and it has been recognized as a significant dilemma for over three hundred years. The issue is the undeniable fact that we each have a direct experience of the content of our own consciousness. We know 'what it is like' to be us, to have our experiences, to think our thoughts, and feel our feelings. Further this sense of self-awareness carries with it the distinct feeling of an *inside,* each of us has a rich experience of the tremendous depths of our own subjectivity. But from a perspective that sees everything as material and objective, the experiential reality of subjectivity is hard to explain... The very idea that the universe is purely material, that all phenomena can be explained by or reduced to the laws of physics, is itself highly metaphysical because it is ultimately a proposition and must be taken on faith" (McIntosh, 2012, p. 10). As much as materialist science abhors the sloppy rationality associated with religion, it has itself adopted a belief system, a credo, based on faith not fact!

All of science hinges on observation first, on first-person, subjective experience. Yet none of that is considered in its solely third-person, objective bias. Wallace notes that "Cognitive science has all along been approaching its subject matter from a paradoxical perspective, relying on first-person observations by the scientist, but not explicitly incorporating them into the scientific method" (Wallace, 2008, p. 82). He also reminds us that "Many of the spiritual and philosophical bases of thought in Asia begin with the mind itself, sometimes to the extent that—completely opposite to the view of the West—it is the mind that is the source of the physical universe rather than the other way around" (Wallace, 2008, p. 112).

Hoffman, in his *Atlantic* magazine interview, succinctly reiterates and clarifies the conclusion of quantum dynamics: "The neuroscientists are saying, 'We don't need to invoke those kind of quantum processes, we don't need quantum wave functions collapsing inside neurons, we can just use classical physics to describe processes in the brain.'"

Hoffman counters, noting: "I'm emphasizing the larger lesson of quantum mechanics: Neurons, brains, space ... *these are just symbols we use, they're not real.* It's not that there's a classical brain that does some quantum magic. It's that there's no brain! *Quantum mechanics says that classical objects — including brains — don't exist* [emphasis added].

"So this is a far more radical claim about the nature of reality and does not involve the brain pulling off some tricky quantum computation" (Gefter, "The case against," April 4, 2016). Materialist science seems to want to insist on using a physical paradigm to make sense of a decidedly non-physical process, like trying to use vision to comprehend sounds in the night.

We can note, too, that all of quantum theory, as well as Newtonian physics and relativity theory and a multitude of scientific disciplines, issue from the language of mathematics. Language as well, like mathematics, is symbolic—they are virtual representations of what we experience as the physical world, but their very existence occurs only in our consciousness. Physicist Eugene Wigner pays homage to that fact when he says, "...the enormous usefulness of mathematics in the natural sciences is something bordering on the mysterious, and... there is no rational explanation for it...the miracle of the appropriateness of the language of mathematics for the formulation of the laws of physics is a wonderful gift which we neither understand nor deserve" (Rolston, 2010, p. 5).

It is mysterious because it falls into the realm of consciousness, which is entirely mysterious to materialistic science. Rolston says, "Mathematics is, above all, mental; it is the logical creation of the human mind, and the fact that mathematics repeatedly helps us to understand the structure of the physical world corroborates the belief that the world we inhabit is a creation of mind" (Rolston, 2010, p. 6).

Bottom line: "I'm claiming that experiences are the real coin of the realm," says Hoffman. "The experiences of everyday life—my real feeling of a headache, my real taste of chocolate—that really is the ultimate nature of reality" (Gefter, "The case against," April 4, 2016).

Necessarily, if we are going to take the data from quantum theory seriously, we have no option but to conclude that the material world is illusory, that our experiences are subjective—the coin of the realm—and that matter ensues from mind. We will be compelled to acknowledge that the fictional myth of our physical experience is only that: a fiction and a myth—"an appearance," as Schrödinger said, not the reality: physicality ain't reality!

"When I open my eyes and I have a conscious experience that I described as a red tomato a meter away, I am interacting with some reality. But that reality is not a red tomato and it's not in space and time," affirms Hoffman (Hoffman, 2017, *Simulation*).

So, mind informs the brain. Perhaps the brain is more like a radio, receiving and outputting vital, intelligent, deliberate information from the implicate order, from the quantum mind that is the universe. Via our observation, our consciousness, we are processing that information, and participating as free agents in co-creating this moment now.

ARTIFICIAL INTELLIGENCE

Can artificial intelligence technology ever arrive at mind? Lloyd thinks so, because he regards information as physical. He thinks of the universe as a quantum computer—"thinking and alive." However, in *Programming the Universe,* Lloyd, in contrast to the premises of quantum realism, particularly the notion of the primacy of consciousness, starts with the premise "information is physical:" "All information that exists is registered by physical systems, and all physical systems register information" he states (Lloyd, 2006, p. 213).[95] And so, for Lloyd, information processing is data computation, calculations involving numbers or quantities in physical systems. But calculation is different than thought, which according to Merriam-Webster is "the action of using

---

95   Yes, and the "place" that it registers is non-physical consciousness.

your mind to produce ideas, decisions, memories, etc.," far more than computing and calculating, and something a computer, being programmed externally, could never "organically" accomplish. As Rolston states, "Information is a richer category than computing" (Rolston, 2010, p.10).

For Lloyd a *universal* quantum computer is "caused" by the quantum fluctuations that foreshadowed time and space, its first cause. But what he describes as the "thinking" and "living" capacities of the universe are limited to computational information and data processing—a stark view of thought and life—*unless* he wants to say a whole lot more about the "infinitesimal vacuum" and "underlying quantum fields." Further he regards "the universe [as] *indistinguishable* from a quantum computer" (Lloyd, 2006, p. 54). So who constructed and programmed it? If artificial intelligence (AI) simply means computational information and data processing, that kind of "consciousness" already exists and will likely evolve over time.

We are suggesting something beyond Lloyd's conceptions of thought and aliveness. Guth's theory of "random underlying quantum fluctuations in an infinitesimal vacuum" represents more than computation and calculation: we are going beyond computational "thinking" and data processing toward the richer dimensions of thought that are hinted at in the dictionary.

Lanza writes, "Until we understand the nature of consciousness, the machine can never be made to duplicate the mind of a man, or pigeon, or even of a dragonfly. For an object—machine, computer—there is no other principle but physics. In fact it is only in the consciousness of the observer that they exist at all in time and space. Unlike a man or pigeon, they do not have the unitary sense of experience necessary for perception and self-awareness, for this must occur before the understanding generates the spatio-temporal relationships involved in every sense experience, before the relationship between consciousness and the spatial world is established" (Lanza, 2009, p. 176).

Harvey concurs: "Machines might replicate many of the intellectual computation abilities of mind-brains, but consciousness remains elusive" (Harvey, 2005, p. 192).

In an interview with Penrose, Steve Paulson says that Penrose "believes we must go beyond neuroscience and data processing, and into the mysterious world of quantum mechanics to explain our rich mental life...

'Most scientists regard quantum mechanics as irrelevant to our understanding of how the brain works...' says Penrose. 'Artificial intelligence experts have been predicting some sort of computer brain for decades, with little to show so far. And for all the recent advances in neurobiology, we seem no closer to solving the mind-brain problem than we were a century ago. Even if the human brain's neurons, synapses and neurotransmitters could be completely mapped—which would be one of the great triumphs in the history of science—it's not clear that we'd be any closer to explaining how this three-pound mass of wet tissue generates the immaterial world of our thoughts and feelings. Something seems to be missing in current theories of consciousness'" (Paulson, "An interview with," May 4, 2017).

Deepak Chopra, in his article, "Artificial Intelligence Will Never Rival the Deep Complexity of the Human Mind," weighs in: "One branch of AI believes that a computer will one day duplicate how the human brain works, once the technical difficulties are worked out...

"AI has taken us to the verge of an Orwellian dilemma, because the spectacular advantages offered by computers weigh so heavily and create such enormous optimism, it's easy to overlook one flaw: AI isn't based on the truth. Computers process information at lightning speed and their abilities improve as the algorithms that are programmed into them become more sophisticated. Yet, without question, life isn't algorithmic, which means that no computer can ever truly be alive. Computers cannot and will never have minds...

"This assertion runs contrary to every part of the AI worldview, which not only foresees mind-like computers but also, in one extreme fantasy, declares that a human being can be digitized, placing every memory into computer memory, along with a lifetime's worth of experience—thus creating a viable afterlife. A digitized human being would be equal to a living human according to this fantasy. There would be no need for a body when life comes down to nothing except information" (Chopra, "Artificial intelligence," Dec 06, 2017).

Underlying this entire discussion, let's note once more, in this

Materialist Paradigm, that the reality of consciousness has not been mentioned or explored. What might artificial intelligence mean without it?

In a world where matter is primary, of course, one could envision that humanity might have the capacity to take on any material challenge successfully. But this brand of science—discounting what quantum dynamics has to say— knows little about the phenomenon of consciousness and how it shapes and informs our world and hasn't begun to wrestle with true intelligence—the authentic kind, not the artificial form. Without entertaining the much more plausible notion that indeed, it's consciousness first, only false assumptions can inform—and so derail—any hope that AI has of grasping the magnitude of mind.

Chopra identifies "five directions for knowing the limitations of computers as thinkers:

1. Computers can calculate anything but understand nothing,
2. Computers cannot truly create — only recombine what humans create,
3. Computers are strictly rational. A human mind owes its richness largely to non-rational aspects,
4. Computers have no insight. They are immune to "aha" moments, and
5. Computers cannot relate to human existence at levels we most cherish — love, beauty, truth...

"The mind isn't the same as the brain," he says, "Therefore, no matter how perfectly one duplicates the machinery of the brain — replicating biology into hard-wired technology — the end result would be a machine, not a mind." Notwithstanding Siri's responsiveness, "relatability," and intelligence, she's a lousy date.

Pointing to the apparent chasm between what information processing means for materialists, and what information processing means to idealists, Chopra observes, "Of all the weird contradictions that plague modern life, the strangest is the collapse of philosophy with the triumph of science. Aesthetics, morals, love, transcendence, idealism — all of these fields of thought, having persisted for thousands of years, in the East and in the West, mean nothing in scientific terms because they cannot be reduced to data, measurements and experimentation" (Chopra, December 6, 2017, "Artificial").

Given that our world is action, movement, events, and not matter or things, and given that everything we perceive is subjective, we can affirm again, then, that there is no other conclusion to be drawn but that our world is virtual, that what we see is an "appearance," as Schrödinger again has attested: "What we observe as material bodies and forces are nothing but shapes and variations in the structure of space. Particles are just *schaumkommen* (appearances)." We live in a world premised on the primacy of consciousness, not matter.

Imagine—just for fun—that the brain is merely a receiver of information quanta from the quantum mind, or from the implicate order. There would be physiological and measurable processes that we could note, as we could note the physical actions that exist in a radio when it receives radio waves. But those process and actions are not the same as the information being transmitted, like a TV image is not a real physical entity. It is only a measuring device, different from what is being measured. The human brain, in this imagining, would be a competent receiver But an AI brain is as likely to ever be such a receiver, (perhaps because of the ingredient we call life), than is a bicycle.

Wallace discerns, "Since it reflects all the richness and complexity of our existence, we tend to identify mind with its reflections rather than the awareness that makes them visible" (Wallace, 2008, pp. 111-112). We identify with what is transmitted, not appreciating that the reality emanates from the transmission.

Consciousness is the canvas of our existence. It is our existence. Without our conscious perception, as Lanza says, there is nothing. Perception *is* existence. These are not considerations in a Materialist Paradigm: the Materialist Paradigm, we have seen, sees only the physical world. The physical world has a reality, but it is, finally, an illusion. Beyond this physical world we see is a deeper reality, an emergent reality, but we have to want to see it.

Chapter Six:

$\sim\!\!\sim\!\!\sim$

AN ALTERNATIVE CONJECTURE— THE UNIVERSE IS VIRTUAL

REALITY IS THE OBSERVER OBSERVING

"It is true that consciousness is the mere surface of our minds, which, as of the Earth, we know only the crust. Below the level of conscious thought, we can conceive unconscious neural states. But these mental faculties, in themselves, apart from their relation to our consciousness, cannot be said to exist in space and time, any more than does a rock or tree."

—Robert Lanza (Lanza, 2009, p. 182).

"Paramount in any effort to comprehend reality at a deeper level is to understand that what you have been witnessing since before you were born—all that stuff 'out there'—has actually been an internal model: a construct within mind that we presume represents something that should be 'out there.'"

—Eben Alexander (Alexander, 2017, p. 53).

As the forces of scientific orthodoxy insist stubbornly that only that which is material is real, the theory of quantum dynamics just as persistently and far more graciously illustrates that, at the very least, there remains something bewildering and insufficiently scrutinized— and apparently contentious—to be considered. The quantum theory concludes that reality, however we might perceive it, does not exist until we observe it.[96] That is, "physical materiality," itself illusory, doesn't exist until we observe it. This is the central tenet of quantum. This *must* mean, it seems to me, that logically and reasonably, consciousness—mind[97]—is required for what we consider the material world to exist. Consciousness first.[98] Or as Walker states, "reality is the observer observing" (Walker, 2000, p. 309).

Penrose reflected that, "It is a striking fact that almost all the interpretations of quantum mechanics...depend to some degree on the presence of consciousness for providing the 'observer' that is *required* [emphasis added] for...emergence of the classical-like world" (Rosenblum and Kuttner, 2011, p. 203).

Any efforts to make quantum dynamics fit with classical dynamics and any efforts to describe quantum as ratifying the primacy of matter are ill-founded and simply backwards. Quantum observation is primary. So if we try to understand the dynamics of quantum—assuming that the material world flows from quantum reality, *not* quantum from material—how then do the features of quantum dynamics— of these mind-bending conclusions we've described—perhaps begin to make sense?

Whitworth states that he believes that "the error of physical realism … contradicts both facts and logic…" asking, "How can a world with random events be also determined? How can a physical universe complete in itself also begin in a big bang? A physics based on illogic builds

---

96 The universe, through us—and likely through other living beings and perhaps even electrons— observes itself. The chicken/egg question arises: how are we here to observe if it takes observation for us to be here? And that's what quantum says! Booting up a virtual reality, akin to turning on a TV, and thus initiating the program, appears to be one of the most likely common-sense ways we can conceive of as to how it all might work.

97 Just to be clear, I am using the dictionary.com definition of mind, (which is consistent with other definitions): "the totality of conscious and unconscious mental processes and activities." This definition incorporates more than merely personal or even human mental processes and activities, and is thus consistent with the idea that all mind is one, or all consciousness is one.

98 ...and in the end, or, rather, in the beginning, that consciousness had to be that of the booter-uper.

paradox into its foundations" (Whitworth, 2014, *Quantum*).

Whitworth (and other scientists, researchers and philosophers we have introduced here), have suggested that quantum mechanics makes much more sense, and it is more consistent, if we begin with the assumption of the primacy of consciousness—*the primacy of the quantum field itself*—rather than assuming material primacy, that matter appeared first in the universe.[99] Even the big bang and inflationary theories of the beginnings of our universe declare that it is the *quantum fluctuations* that were present in the *infinitesimal vacuum* which spawned the big bang. It is quantum manifestation which germinated physical manifestation. Quantum dynamics are demonstrably non-physical dynamics, as we have seen. So, if quantum dynamics—the "quantum fluctuations in the infinitesimal vacuum" out of which the big bang emerged according to orthodox science—are not physical, how then could physical be primary?

Whitworth continues, "The honest answer is that [real objects] cannot [be constituted from unreal components], and to invent a mystical wave-particle duality to cover this up is to institutionalize illogic. It *must* be that quantum events create physical events, *which can only be if the physical world is virtual*" [emphases added]).[100] Quantum theory says that observation creates reality, and as "hard" and physical as "material reality" seems, according to the theory, it cannot exist. Planck's assertion again is: "As a man who has devoted his whole life to the most clear-headed science, to the study of matter, I can tell you that as a result of my research from the atoms, this much: There is no matter as such!"

The physicality we observe, according to quantum theory, does not exist. So what does?

---

99   Matter cannot have appeared first, as it is the underlying quantum reality—according to the data gleaned from traditional scientific lexicon—which generates what appears to us to be solid material, but which is actually quanta: "packets of action."

100 Retrieved from the website: "The **definition of virtual reality** comes, naturally, from the definitions for both 'virtual' and 'reality'. The definition of 'virtual' is near and 'reality' is what we experience as human beings. So the term 'virtual reality' basically means 'near-reality'. This could, of course, mean anything but it usually refers to a specific type of reality emulation…

"Our entire experience of reality is simply a combination of sensory information and our brains sense-making mechanisms for that information. It stands to reason then, that if you can present your senses with made-up information, your perception of reality would also change in response to it. You would be presented with a version of reality that isn't really there, but from your perspective it would be perceived as real. Something we would refer to as a *virtual reality*…." (Virtual Reality Society, "What is," 2017). https://www.vrs.org.uk/virtual-reality/what-is-virtual-reality.html.

VIRTUAL REALITY

"The day science begins to study non-physical phenomena, it will make more progress in one decade than in all the previous centuries of its existence. To understand the true nature of the universe, one must think in terms of energy, frequency and vibration."

– Nikola Tesla (Walia, March 8, 2014, "Science proves").

"The odds we are in base [i.e. not simulated] reality are one in billions."

—Elon Musk (Musk, June 2016, Code Conference).

"For the moment, therefore, we'll accept on a provisional level that what we clearly and unambiguously recognized as existence must begin with life and perception... Once one fully understands that there is no independent external universe outside of biological existence, the rest more or less falls into place."

—Robert Lanza (Lanza, 2009, pp. 16-17).

A word on perspective: As we explore this subject, let's not lose sight of what we have already established: that, as I noted, "We exist in a living universe (and on a living Earth). The cosmos is an alive, intelligent, conscious, self-reflective, and relational unfolding. We are involved in an emerging, evolving, intentionally creative process of being and becoming." All of this is true within the virtual game we are playing which we believe to be physical reality. And in the game, Lanza's quotation here makes sense. In the game is only biological existence. We will look at that subject in detail in Chapters Seven, Eight, and Nine. Said otherwise, the Consciousness Paradigm is a virtual reality paradigm.

Film, television, and digital media have been around for a while, and are all examples of VR technology. Technology, however, leapt forward with the advent of 3D VR "games" that capture sight, sound, direction, and even scents, allowing the player to have multisensory and deeply absorbing experiences. "Virtual reality (VR) has been around since the 1980s," notes Michael Greene, referring to this technology of virtual "games." He comments, "Watching people's reactions when they try on VR headsets for the first time reminds me of the stories about the audience reactions at the dawn of the movie era. Legend has it that mayhem broke out at the theater in 1896 when the Lumière brothers presented their one minute documentary, 'Arrival of the Train at La Ciotat.'[101] Incredibly basic by current standards, the film simply showed a train coming into a station from the point of view of someone standing right along the tracks. But never having ever seen anything like it before, people jumped out of their seats, terrified that they were about to get run over. Of course, as years have passed, we have become accustomed to the marvels of film techniques, and more and more able to maintain some sense of distance, mindful that we are not actually *in* the film. VR, on the other hand challenges some of our physical and emotional boundaries, altering our immediate experience of what's real and blurring our sense of being separate from what we are watching."

Similar in novelty to the unfamiliarity of a first exposure to film, especially one such as "La Ciotat," must be one's first exposure to virtual reality in the form of VR goggles. Greene writes: "Put on your VR goggles and headphones and you enter a new environment, experienced from a first-person, 360-degree perspective. Like real life, when you turn your head to the right or look up, the sights you see and sounds you hear are different than when you look down or turn toward your left. While you know you're wearing goggles and headphones, you feel as if you've been transported into some other world. It's immediate and it's immersive" (Greene, M. *Psychotherapy Networker,* November-December 2016, "Is VR" p. 34).

When we play a VR game, we opt to suspend our everyday experience and enter a fantasy world that seems persuasively "real," though we know that it isn't. In the virtual reality conjecture, suspending our

---

101 From the You Tube page "Arrival of a Train at La Ciotat (The Lumière Brothers, 1895)." "Original Title: L'Arrivée d'un Train Á la Ciotat." Directors: Auguste and Louis Lumière Year: 1895 The first public exhibition of motion pictures occurred on 28th December 1895 when August Lumière and Louis Lumière (the Lumière Brothers) exhibited a selection of ten of their single-reel films to a paying audience at a Parisian cafe. 'Arrival of a Train at La Ciotat' is considered to be the first motion picture in modern history (although more an experiment from the Lumière-brothers to use their 'invention' of film, it shows a train arriving at a passenger station). Popular legend has it that, when this film was shown, the first-night audience fled the cafe in terror, fearing being run over by the "approaching" train. Most of the cast were members of the Lumière family and employees from the Lumière factory." https://www.youtube.com/watch?v=1dgLEDdFddk

A Universe Full of Magical Things

everyday reality allows us to see that *that* experience is the fantasy, and that it is *our everyday experience that is symbolic, and not real.* Our everyday experience *is* a virtual game.

So to pursue a virtual (quantum realism) conjecture[102] requires we set aside for the moment the assumptions we've made, conclusions we've drawn, and things we've been told about what the universe is and how it functions—because if we start with the familiar premises our culture promotes, we'll end up with the same perplexity about quantum theory, and the same mistaken view of our existence.

Instead of these familiar, yet inadequate and indefensible conclusions our culture and its science asks us to accept, consider instead the following:

- We exist in a material-plane-virtual-game, wherein what we experience as physical is only real within the game.
- There is beyond the game nothing physical, objective, deterministic, or predictable.
- Matter itself does not exist. What we experience as matter and the entire material plane within which we seem to exist is illusory, virtually "real" and "true" only in the context of our perceptions: "The only things we perceive are our perceptions," as the philosopher George Berkeley observed.
- This virtual reality, this ultimately non-physical "reality," this game, was birthed out of non-physical "quantum fluctuations in the infinitesimal vacuum," i.e., quantum mind, or

---

102 "Virtual reality entails presenting our senses with a computer generated virtual environment that we can explore in some fashion…Virtual reality is the term used to describe **a three-dimensional, computer generated environment** which can be explored and interacted with by a person. That person becomes part of this virtual world or is immersed within this environment and whilst there, is able to manipulate objects or perform a series of actions. Virtual reality technology needs to take our physiology into account…

If an implementation of virtual reality manages to get the combination of hardware, software and sensory synchronicity just right it achieves something known as a *sense of presence*, where the subject really feels like they are present in that environment…

There are many different types of virtual reality systems but they **all share the same characteristics** such as the ability to allow the person to **view three-dimensional images**. These images appear life-sized to the person. Plus they change as the person moves around their environment which corresponds with the change in their field of vision. The aim is for a seamless join between the person's head and eye movements and the appropriate response, e.g. change in perception. This ensures that the virtual environment is both realistic and enjoyable." (Retrieved from the Virtual Reality Society website: 2017. https://www.vrs.org.uk/virtual-reality/what-is-virtual-reality.html).

consciousness itself.

- In this virtual game, all that we know we know subjectively—uniquely, from our own point of view. In this game, time and space and matter/energy are apparent only for us, as observing/participating biological beings. In the game, life and mind have existed in the universe since the time of the big bang or metaverse.

- Such notions as "infinity" are meaningless, because there is nothing physical that might or might not be "bounded." Physicality in a virtual game is a dream or an illusion, pretend, symbolic, and virtual: a fantasy concept.

- Our scientific studies and explorations and experiments and conclusions are applicable, but only in the virtual game, this material-plane-fantasy. Our conclusions about physicality and materiality are valid only in the game, just as would be our conclusions in the context of any other virtual reality game we are familiar with.

- What is real *beyond* the game is consciousness, in the "form" of a creative "programmer" who booted up this game, along with any context that involves such a programmer. *In* the game all that is real, all that has any meaning, has to do with our individual points of consciousness, observing the game, and further, participating with the game—with the universe—in what next evolves, as though we had free will.[103]

---

103 Lanza notes: "…Recent experiments by (researcher Benjamin) Libet, announced in 2008, analyzing separate, higher-order brain functions, have allowed his research team to predict *up to ten seconds in advance* which hand a subject is about to decide to raise. Ten seconds is nearly an eternity when it comes to cognitive decisions, and yet a person's eventual decision could be seen on brain scans that long before the subject was even remotely aware of having made any decision. This and other experiments prove that the brain makes its own decisions on a subconscious level, and people only feel that 'they' have made and performed a conscious decision. It means that we go through life thinking that, unlike the blessedly autonomous operations of the heart and kidneys, a lever pulling 'me' is in charge of the brain's workings. Libet concluded that the sense of personal free will arises solely from a habitual retrospective perspective of the ongoing flow of brain events…

"What, then, do we make of all this? … We can only see our own mind or, perhaps, it's better put that there is no true disconnect between external and internal. Instead, we can label all cognition as an amalgam of our experiential cells in whatever energy field may pervade the cosmos" (Lanza, 2009, p. 38-39).

"In a controversial 2011 report entitled 'Feeling the Future,' psychologist Daryl Bem of Cornell University presented compelling evidence of precognition—that is, that people demonstrate conscious cognitive awareness of impending stimuli *before* the computer has randomly selected the

-   We participate in the game until we don't. We might presume that the part of us that is observing—consciousness—may have a continuing essence beyond the game. The Dutch Enlightenment philosopher Benedict Spinoza expressed this sentiment thus: "The human mind cannot be absolutely destroyed with the human body, but there is some part of it which remains eternal" (Spinoza, *The Ethics,* 1677). Lanza's biocentric point of view "of the timeless, spaceless cosmos of consciousness allows for no true death in any real sense. When a body dies, it does so not in the random billiard-ball matrix but in the all-is-still-inescapably-life matrix," says Lanza (Lanza, 2009, p. 188).

These contentions represent the virtual reality viewpoint, and as we see, they stand in stark contrast to the materialist contentions our traditional science has asked us all to believe. But these virtual reality contentions directly, consistently, and comprehensively illustrate the uncontested quantum reality convincingly, whereas traditional mechanistic beliefs simply don't know what to do with quantum mechanics.

So instead, let's consider these alternative notions, and see where they take us. We can conceive of these next sections as thought experiments of a kind. What if each of the statements were true? What would be the implications and what would it mean for humanity? Taken together, are they at all persuasive for us? Is it possible that, indeed, what we experience is not a hard, but rather a virtual reality?

Assume that the Quantum Field is Mind

Let's, for now, assume that the Quantum Field is Mind—consciousness. We've already seen that this is the conclusion reached by

---

stimuli to present(!) A follow-up meta-analysis, rigorously constructed from ninety experiments in thirty-three different laboratories spread across fourteen countries, confirmed this experimental violation of the most fundamental notions of materialist science, and our commonsense ideas about cause and effect and the nature of time itself." (Daryl J. Bem, "Feeling the Future: Experimental Evidence for Anomalous Retroactive Influences on Cognition and Affect," *Journal of Personality and Social Psychology* 100, no. 3 (2011): 407-25; and D. Bem et al, "Feeling the Future: A Meta-Analysis of 90 Experiments on the Anomalous Anticipation of Random Future Events, "*F1000Research* 4 (2015), 1188. Doi:10.12688/f1000research.7177.1. (Alexander, 2017, p. 44).

We experience choice as we do space, time, and matter/energy, even though outside the game, none of these exist.

many highly respected scientists and physicists, including Schrödinger, Eddington, Bohm, Einstein, Planck, Walker, and many others. One of several ways we might define mind (i.e. consciousness) is this one, from Larry Dossey, former Chief of Staff at Medical City Dallas Hospital, internal medicine physician, and author of 12 books, who considers "mind" as consisting of all minds, the "one mind." He says that "One Mind is a collective, unitary domain of intelligence, of which all individual minds are a part" (Dossey, 2014, p. xviii).

Dossey also references K. Ramakrishna Rao, (Rao, 2011, p. 352), a contemporary philosopher and consciousness researcher in India: "Consciousness in the Indian tradition…is a fundamental principle that underlies all knowing and being…the cognitive structure does not generate consciousness, it simply reflects it…Consciousness is the light which illumines those things on which it shines" (Dossey, 2014, p. xviii).

In *The Physics of Consciousness,* physicist Walker unabashedly declares: "It is Consciousness that began everything, that grows matter into a universe of existence; it is Consciousness that unifies and constrains all of us as individual beings; it is Consciousness that orders space and time out of a chaos of random events" (Walker, 2000, p. 334). This perspective on our existence results in a paradigm where consciousness is primordial, a paradigm that completely shifts our values and motives for living away from a matter-based frame of reference.

Walker concludes, definitively, from his work, as articulated in that book, that "We have found, above all else, that *quantum states and mind are one and the same thing* [emphasis added]. We have found that in their essential nature, quantum fluctuations are the stuff of consciousness and will. And now, here, we find that this mind stuff was the beginning point of the universe—the stuff that out of a formless void created everything that was created. We find that in the beginning there was this quantum potentiality. We discover that in the beginning, there was the Quantum Mind, a first cause, itself time-independent and nonlocal that created space-time and matter/energy" (Walker, 2000, p. 326). This statement summarizes and differentiates the alternative point of view we are suggesting from the familiar materialist one.

So, again, in the beginning, in the appearance of an electron, mind was somehow present. Recall Dyson: "It appears that mind, as manifested by the capacity to make choices, is to some extent inherent in every electron" (Frankenberry, 2008, p. 372).

The underlying quantum reality is not physical, as we've come to realize. It is however, as Lloyd and others have described, an action of "information processing"—but information beyond Lloyd's computational "thinking" and data processing "aliveness." It is information processing that is "the action of using…mind to produce ideas, decisions, memories, etc." It is an aliveness, that is, agency—choice itself: the "capacity, condition, or state of acting or of exerting power," and it is *intentionality!* Of course, it takes non-physical mind to process that kind of information, to create out of that kind of information. Ultimately, it is quantum mind which processes information, which thinks. Which is conscious. Which knows.

If we can conceive that consciousness, mind—quantum mind—is both first cause and "the ground of being," that out of which all else arises, we find that quantum dynamics turns out to be the way mind operates and interfaces with materiality. How could it be otherwise, if *"There is no matter as such?"*

Assume that Whatever the First Cause Is Has to Do with Quantum Consciousness

So, further, let's assume for now, that whatever the first cause is has to do with quantum consciousness, i.e. "quantum fluctuations in the infinitesimal vacuum." That conclusion *is* the conclusion that is being arrived at by contemporary science. Recall, Lloyd wrote quantum fluctuations generated the big bang event. Further, he declared, "Quantum mechanics supplies the universe with…random quantum fluctuations… "The universe has systematically *processed and amplified* the bits of *information* [emphases added] embodied in those quantum fluctuations. The results of this information processing is the diverse, information-packed universe we see around us: programmed by quanta…" (Lloyd, 2006, p.61). He states that, "Quantum mechanics describes energy in terms of quantum fields, a kind of underlying fabric of the universe" (Lloyd, 2006, p. 40). That is as far as orthodox science goes. There is no first cause acknowledged or described. However, there *are* "quantum fluctuations in the infinitesimal vacuum," as Guth found and as contemporary science holds, that apparently underlie the universe, and generate the time-space and matter/energy of the universe.

So if the universe has a start, a first cause, the name for that first cause *for science,* would be "quantum fluctuations;" *for religion* that first cause would be called "God;" and *for* those who see *a living, intelligent and interconnected universe*, it would be called "Quantum Mind," "Consciousness," Spirit," or "One Mind."

It seems that whatever potential exists implies an ineffable and enigmatic complexity, "whereof we cannot speak, thereof we should remain silent," except perhaps, circling back, to acknowledge that this physical experience we are having must flow out of the quantum mind.

Whitworth offers this baffled query: "A universe that began either existed before its creation to make itself, which is impossible, or it was made by something else. It is impossible that a complete universe began by itself, from nothing. Yet oddly enough, this is what most physicists today believe. They suggest the first event was a quantum fluctuation of the vacuum… But if matter just popped out of space, what did space pop out of? How can a quantum fluctuation in space create space? How can time itself begin?" (Whitworth, 2014, *Quantum*).

Acknowledge that Materialist Science Has Not and Cannot Address Causality

So, let's, for now, acknowledge that materialist science has not and cannot address causality—because the quantum reality supersedes cause and effect materialist reality. In materialist reality, everything is tied to a cause and effect process. However, in quantum theory, there is only uncaused or self-caused reality. It is, finally, quantum reality that "is the underlying fabric of the universe." Quantum reality, we have recognized, is indeterminate, unpredictable, and random for us. This fact is what materialist-biased quantum interpretations want to, have not, and cannot, explain.

This underlying fabric is mind, is conscious: quantum says that for the universe to exist requires observation. And materialist science is again without any answers.

Acknowledge That a Boot-Up of Our World in an Instant
Is More Credible Than Something Out of Nothing

Let's, for now, acknowledge that a boot-up[104] of our world in an instant is more credible than "something out of nothing," the "no-space" and "no-time" conjecture, considered to have preceded the big bang. The metaphor (or is it?) of a virtual game—a virtual world—starting up when a switch is flipped or a button pushed is a scenario more aligned with quantum theory, and may lead to more useful insights than the materialist non-answers can provide.[105]

The fact is that a materialist science of *cause and effect reality* asks us to believe that the universe suddenly showed up *without* cause! Of course this idea has not and likely could never be documented. It is an article of faith. Science's resistance to the idea that there could be a first cause leaves no other option except, ironically, to understand that the beginning of our universe is a miracle. There is no explanation for how "quantum fluctuations in an infinitesimal vacuum" could arise out of no time and no space. That the universe just "showed up," as science seems to be declaring, is incomprehensible to us humans: it is miraculous.

Or, alternatively, the universe was conceived of and initiated by some "superintellect" (Hoyle), "organizing principle," (Sandage), "universal mind," (Eddington), "supernatural agency," (Greenstein) or "conscious and intelligent mind," (Planck), a notion so extreme for mainstream scientists as to be considered miraculous as well.

Once we agree to let ourselves to ponder a first cause, we see that there are many possibilities that have emerged from the human mind to conceive of what that first cause might be. The standard model says that the universe had a beginning, although no explanation of how or exactly what is given. László's notion of an evolving Metaverse suddenly giving birth to our universe is another possibility, although it

---

104 When a computer is off, there is nothing about it that serves. When it is turned on, or booted up, an unending plethora of stories and imagery, in a moment, comes into being. A virtual reality conception of existence presupposes a "programmer" of some sort who first designed the computer, and then turned it on, i.e., booted it up.

105 Quantum theory verifies that matter is illusory, "not real," but rather, packets of action. Curiously, "illusory" is a synonym of the term "virtual." Other synonyms include: simulated, artificial, imitation, and make-believe.

merely relocates the problem of first cause. A virtual reality booted-up in the form of a big bang—like a game or a computer or TV, allows us to make better sense of what quantum mechanics tells us is true about our reality, is more consistent, and permits us to draw more meaningful conclusions, as we are seeing.

Whitworth, who holds that the universe may just be a virtual game or virtual computer, or a virtual reality, (a conjecture which is, by inference, a first cause contender), states that "if the physical world is a virtual reality, the big bang was the system boot up" (Whitworth, 2014, *Quantum*). The big bang turns on the physical world, or, alternately, the booter-uper flips the switch, and virtual reality emerges.

In the virtual reality conjecture, some "first cause" is implicit. Call it "quantum fluctuations in the infinitesimal vacuum," God, or Quantum Mind, but our existence is the result of a deliberate, intentional, designed creation at the hands of some transcendent essence, some all-pervasive presence.

We are a kind of virtual game. If not, then we have to entertain what I would call "the cherry pie conjecture"—that from an empty oven, suddenly, out of nothing, and without any explanation, a warm, enticing cherry pie, fully and freshly baked, appeared, ready to be devoured.

We get to choose between miracles: there is no third option.

Because quantum theory is authoritatively considered to be the base of our reality, and that base is not material but action and movement, we must conclude that the "hardness" of material experience, matter itself, is an illusion. When we play a virtual game, we relate to the avatars and "characters," the "physical" location, the "relationships," and the events as if they are real. We know that they are real for us in the game, but that what is real ultimately *is not the game at all.*

The same is true of the picture the pixels create on our TV screen: the picture is virtual, not real. When we watch a movie, we are watching a series of digital pictures, frames, themselves an assemblage of pixels, which pass our field of vision in rapid sequence, appearing real, appearing to have a story—yet, in reality, we are only watching a representation, a virtual reality. We understand that the game, the TV, or the movie are "pretend," but for the moment we relate to them *as though they were real.*

Quantum realism, or virtual reality, posits that our everyday lived experience is not what's real, but a representation of reality—like an icon on a computer has meaning for us as a representation of some action we access through clicking on the icon. The icon is only symbolic. Our lived experience is only symbolic, a virtual game we are playing. A virtual game, a TV screen, or a movie are all only symbolic: they are not reality. So too, our lived experience!

Further, just as an avatar in a virtual game is not invented when the pixels are programmed into the game, but when she is *conceived of in the creative imagination of her inventor,* so too we are not brought into existence on this planet by our mothers but by the creative vision imagined by the quantum mind.

So, if that is so, if we are conscious living characters in an intricate and complex, pre-designed game, what's the point of the game? *Is* there a point?

The answer to that question is personal and subjective, only a perception, and it depends primarily on the worldview we hold and on the values we live based on that worldview. Is the point survival, dominance, subservience, acquiescence, competing, acquiring, cooperating, learning, creating, inventing, sharing, discovering, loving, nurturing, honoring, appreciating, evolving, or something else? We will play the game according to what we value, and what we value is a function of our degree of consciousness, the developmental stage of growth we have come to as we engage in and experience our lives. That's a subject we'll tackle more in Chapter Ten.

In any case, we can see that a virtual reality conjecture reaches a conclusion about our existence itself: it posits a first cause and it posits that there is therefore, somehow, a purpose for our existence. It is a conjecture that, additionally, is highly congruent with the principles and laws of quantum dynamics. As it is based on the proposition of "consciousness first" instead of the materialist assumption of "matter first," quantum reality resolves the seeming incompatibilities of Newtonian physics with the data of quantum dynamics—because matter is understood to be virtual, not real, so we need not treat the material world as a concrete reality. It is only an abstraction of what is real.

In a virtual game, infinities are irrelevant and make no sense: the game goes off in whatever direction and however far a player chooses to act. There are no physical limits—a virtual universe is unbounded because there is no physicality to it. It's a story—nothing physical exists. We pretend that there is a "hard," concrete, physical and material reality. But we know what we are engaging in is a collection of a kind of ever-progressing, ever-unfolding pixels, nothing more and nothing real. Still we play the game as if it were all that is.

The game is how we know we exist. Our being conscious of our "existence" here depends on our having an environment, a world, that is, a game, with which we interact. It depends on an ability to observe.

We play the game as if there were hard, physical materiality. As if there were three dimensional pixels. We play the game as if it were all that is. Yet, in the end, we find, quantum dynamics works because it is at the base of our existence, and because there is no true physical reality, only quantum reality, only consciousness. At the base of our existence is only consciousness, observing and participating in this virtual, nonphysical game.

Consider that Quantum Realism Addresses
Conundrums of Quantum Theory

So let's consider that a virtual reality construct (quantum realism) can address many of the conundrums of quantum theory. In his paper, Whitworth articulates a variety of features of quantum realism and its virtual reality model that readily resonate with quantum theory. For example, he says that "A virtual reality is a world created entirely by information processing and processing is the transformation of information values. As virtual worlds exist by processing, by definition nothing in a virtual world exists independently in or of itself" (Whitworth, 2014, *Quantum*).

Quantum theory is an information processing theory, which pronounces that all that exists are informational probabilities—existing nowhere but in our consciousness, the action of information processing itself—not a thing, but an action. Virtual reality—what we see on a virtual game, or a computer, or on a TV—is only information processing. We are seeing coded data, programmed pixels, symbols pointing toward an underlying reality. But by no means reality itself. The consciousness we experience as we engage in the game is the only part of the game that is not virtual—just as a player in a video game is the only part of the video game that is real. But even our human reality has no physical component. What is "real" is our witnessing and our participating, the actions of consciousness, the living of life.

The Virtual Reality Society construes, again, that "Our entire experience of reality is simply a combination of sensory information and our brains' sense-making mechanisms for that information. It stands to reason then, that if you can present your senses with made-up information, your perception of reality would also change in response to it. You would be presented with a version of reality that isn't really there, but from your perspective it would be perceived as real" (Virtual Reality Society, 2017, "What is").

All of which is similar to the experience of watching a movie projected on a screen or wall. When we watch the screen where the projector has been focused, what we see is real enough for us to relate to and identify with—even though the things we see are only simulated representations. The most recent technological coup is a virtual game player so vivid that players can be seen falling over or banging into things, because what they perceive, via enhanced programmed multi-sensory technology, is not being differentiated from what is "real" beyond what is viewed.

Quantum realism proposes that our physical world gives us symbolic, not real, data, as Hoffman indicated. Whitworth affirms as well: "...our reality is a virtual reality that only exists by information processing beyond itself, upon which it depends."

"In quantum realism," Whitworth states, "the observer is real not the observed, so physical events only occur if observed, as quantum theory says." He explains that "a primal quantum reality generates physical reality as a program creates pixels. So physical systems emerge from a fundamental quantum reality.... [which] consists of dynamic events not static things," (and which describes virtual reality as well). Observation results in pixels, avatars, physical structures, stories, and everything else that is observed, yielding the symbols, representations, and other information that is processed by mind. But information is nothing and does not exist without mind to process it—without an observer to ratify its existence. When we observe a game in a null state, what we are looking at is akin to the superposition concept of quantum dynamics: Like quantum, it is only in observing and participating in the game that we can have access to its information.

"If the processing stops so does the virtual reality," Whitworth adds. "That is, if quantum processes stop, so does our reality." If the TV or the computer or the game goes off, so do those realities. If mind stops, so does our reality.

"'Causality,' such as we perceive it, then arises not from static states but from dynamic events," says Whitworth. "Past, present, and future are not properties of four-dimensional space-time but notions describing how individual IGUSs [information gathering and utilizing systems] process information" (Whitworth, 2014, *Quantum*).

"In a virtual reality model," according to Whitworth, "a space that seems empty to us is actually full of processing. This 'fullness' is the grid network that mediates light, generates vacuum energy and produces the Casimir effect [physical forces arising from a quantized field]. So there isn't a physical ether but there is a non-physical grid [the zero-point field] that mediates all physical events. The quantum grid flawlessly reflects light and matter within itself…" (Whitworth, 2014, *Quantum*).

In quantum theory, "… space, which has so much energy, is full rather than empty," says physicist James Hartle.[106] In the virtual reality model, this fullness of space is null processing, like an idle computer that is actively running a null cycle over and over. So empty space isn't empty,[107] it is the "grid network"—quantum fluctuations. It is mind.

Whitworth affirms, *"Quantum realism is the…view that the quantum world is real and is creating the physical world as a virtual reality. Quantum mechanics thus predicts physical mechanics because it causes them* [emphasis added]."

Additionally he notes, "The observer effect is how the physical world arises at all. The long sought boundary between the classical and

---

106 Whitworth referencing Hartle, J. B. (2005). The Physics of "Now." *American Journal of Physics*, 73, 101–109, avail at http://arxiv.org/abs/gr–qc/0403001, p. 101.

107 Whitworth referencing Bohm, D. (1980). Wholeness and the Implicate Order. New York: Routledge and Kegan Paul, p. 242.

quantum worlds is the 'click' of observation."[108]

Quantum reality exposes materiality as a representational and virtual perception.

Appreciate that, Like In a Virtual Game, What We See Is a Simulated Reality

Let's appreciate then, that, like in a virtual game, what we see is reality simulated, like an icon of a computer isn't the process but seems like it is, or like an avatar is real only for a virtual game and represents something *other than* the pixels our senses perceive and even something *other than* the avatar itself.

All of us make assumptions that allow us to make decisions as to how we want to proceed right now. Amanda Gefner in her 2016 *Atlantic* magazine interview with Donald Hoffman makes this observation: "As we go about our daily lives, we tend to assume that our perceptions—sights, sounds, textures, tastes — are an accurate portrayal of the real world. Sure, when we stop and think about it—or when we find ourselves fooled by a perceptual illusion—we realize with a jolt that what we perceive is never the world directly, but rather our brain's best guess at what that world is like, a kind of internal simulation of an external reality (Gefner, 2016, "The evolutionary").

Gefner says that Hoffman "has spent the past three decades studying perception, artificial intelligence, evolutionary game theory, and the brain, and his conclusion is a dramatic one: The world presented to us by our perceptions is nothing like reality" (Gefner, 2016, "The evolutionary").

Hoffman tells us, "There's a metaphor that's only been available to us in the past 30 or 40 years, and that's the desktop interface. Suppose there's a blue rectangular icon on the lower right corner of your computer's desktop—does that mean that the file itself is blue and rectangular and lives in the lower right corner of your computer? Of course not. But those are the only things that can be asserted about anything on the desktop—it has color, position and shape. Those are the only

---

108   Whitworth referencing Cole, K. C. (2001). The hole in the universe. New York: Harcourt Inc..

categories available to you, and yet *none of them are true about the file itself or anything in the computer.* They couldn't possibly be true. That's an interesting thing. *You could not form a true description of the innards of the computer if your entire view of reality was confined to the desktop* [emphasis added]. And yet the desktop is useful. That blue rectangular icon guides my behavior, and it hides a complex reality that I don't need to know. That's the key idea" (Gefner, 2016, "The evolutionary").

We are wise to take icons seriously if we wish to navigate a computer. But Hoffman warns that we are wise as well to take the icon figuratively, as a symbol, and not literally: "We've been shaped to have perceptions that keep us alive, so we have to take them seriously. If I see something that I think of as a snake, I don't pick it up. If I see a train, I don't step in front of it. I've evolved these symbols to keep me alive, so I have to take them seriously. But *it's a logical flaw to think that if we have to take them seriously, we also have to take them literally* [emphasis added]… Snakes and trains, like the particles of physics, *have no objective, observer-independent features.* The snake I see is a description created by my sensory system to inform me of the fitness consequences of my actions. Evolution shapes acceptable solutions, not optimal ones. A snake is an acceptable solution to the problem of telling me how to act in a situation. My snakes and trains are my mental representations; your snakes and trains are your mental representations" (Gefner, 2016, "The evolutionary"). Again, think of the desktop icons. They are snakes and trains, which, like the desktop icons indicate a reality but *are not* the reality, only a simulation of the underlying reality.

Similarly, the words you read here represent symbols only, concepts and ideas and imagery only, perceptions only. The letters and the words and the punctuation and the order, and the language as well, result in conventional symbols we use to communicate about our perceptions, but surely are not "things" in themselves.

Perceptions are not "objectively" accurate—there is no objectivity in quantum or virtual reality—so they are not the same as any other person's. What we see is, in truth, neither absolute nor objective: "The central lesson of quantum physics is clear." Hoffman attests, "There are no public objects sitting out there in some preexisting space. As the physicist John Wheeler put it, 'Useful as it is under ordinary circumstances to say that the world exists "out there," independent of us, that view can no longer be upheld'" (Gefner, 2016, "The evolutionary"). Robert Lanza and Brian Whitworth would echo that point of view as well.

If we consider what Hoffman is telling us—that our perceptions are no more real than the computer icon is a substantial thing—we are de facto describing a virtual reality. When we perceive what seems to be physical reality, according to quantum mechanics, we are de facto describing a virtual reality!

Whitworth comments, "We only see a construct, an interface to reality, just as an email is an interface to a person" (Whitworth, 2014, *Quantum*). These observations, out of which we interpret and construct meaning, represent, but are not the same as what is real. What's real is the unitary quantum mind with "the capacity to make choices, [which] is to some extent inherent in every electron" (Frankenberry, 2008, p. 372).

Acknowledge That We Can Experience Our Reality as Our Reality, But Not the Reality

Let's, for now, then, acknowledge that, like a virtual game, we can experience our reality as real, and as *our* reality—but not *the* reality. Encoded (physical) reality or icons or TV personalities or virtual avatars and pixels and snakes and even language and mathematics are our reality, but these are only useful symbols of what makes up the real, and only *represent* what is real. They are a simulation only, not underlying reality. The underlying reality is, instead, "a kind of underlying fabric of the universe," as Lloyd says (Lloyd, 2006, p. 40), or "an underlying field of intelligence," as Chopra says (Chopra, 2006, p. 26).

Indeed, in space and time, which we experience and take to be objective, "public objects" are themselves representational of the underlying reality, but are not the underlying reality, as we are seeing. They, too, are symbols: "Space and time are convenient ideas for ordinary life but they don't explain the extraordinary world of physics," says Whitworth (Whitworth, 2014, *Quantum*).

A film projected on a white wall captures our attention, lures us in, and tells us a story. But none of what we are paying attention to in that projection is real. What is real is the light itself, emanating from the far side of the projector, enabling us to see the story and to imagine we are it.

Further, quantum processing is ongoing, like a computer, for example, which is processing when it is computing, and is "conserving" processing even when not computing, null processing. The computer stands in readiness for the next calculation, as it were. "Every physical event is a new creation, and only the conservation of processing maintains the illusion that particles continuously exist," notes Whitworth (Whitworth, 2014, *Quantum*).

Only the continual quantum fluctuations in a living (conscious, intelligent, agentic, and emerging) universe maintain the illusion of physical reality, much like only the pixels maintain the illusion that what we see on a TV seems real. He says, "Thinking imaginary quantum events create real physical events is like thinking a TV show is real and the studio behind it is imaginary" (Whitworth, 2014, *Quantum*).

"Quantum realism offers the alternative that there actually is something beyond the physical world," affirms Whitworth. "It is not a heaven or hell but a quantum world that quantum theory has already mapped and quantum computing is already using" (Whitworth, 2014, *Quantum*).

Appreciate that Virtual Space-Time Could Appear to Those within It As Our Space-Time Does to Us

Consider the possibility that, effectively, "virtual space-time could appear to those within it as our space-time does to us" (Whitworth, 2014, *Quantum*). Whitworth refers to a description from Brian Greene to expand on the idea that a space that defines all space can't curve on itself, and a time that defines all time can't be relative and changing and still be an ultimate feature of the universe: "We see time as a stream carrying all before it and space as a canvas upon which all exists, but a little thought denies this. How can a space that defines all directions itself 'curve,' as Einstein says? How can a time that defines all change, itself change? A time and space that change can't be fundamental... many of today's leading physicists suspect that space and time, although pervasive, may not be truly fundamental."[109]

So, according to Whitworth, "not only did our universe begin, but that its space and time did too." And as we've seen, he prefers to understand how that happened as a booting-up of a virtual reality: us, observing. [110]

HOW DO QUANTUM DYNAMICS AND QUANTUM REALISM ALIGN?

So how do quantum dynamics and quantum realism align? Let's review the conclusions we reached in Chapter 2 about the nature of quantum dynamics, and apply them to a quantum realism context:

---

109 Whitworth referencing Brian Greene: Greene, B. (2004). *The Fabric of the Cosmos.* New York: Vintage Books, p. 471.

110 In our virtual "game," we play by the game's rules some of which, for Whitworth, include such specifics as these:

- "In this model, our three-dimensional space is the inner surface of a four-dimensional hyper-bubble."
- "In quantum realism, a physical event is irreversible because it is a processing reboot."
- "A virtual time may slow down if the processing load increases, but a physical event as a quantum reboot implies the arrow of time."
- "In our computing," Whitworth notes, "a reboot is a processor reloading its programs from scratch, e.g. turning a computer on and off reboots it, and loses any work you are doing, unless you saved it!" (Both Whitworth and Walker, and many others, point out that we actually experience discontinuously. We observe in incremental, discrete micro-seconds of time, "packets of action." We can imagine a movie where each frame is revealed at speed faster than our ability to register and process the advancing frames as discrete. The movie thus appears continuous to us, although it actually consists of a series of [discontinuous] individual pictures, individual frames. So what we see as seeming continuity is, in effect, disconnected: reboot after reboot— no wonder we miss so much!)
- "In quantum realism, our space arises from the grid connection architecture and our time arises from the processing cycles of its nodes."
- "All the fields of the standard model [are] different processing effects."
- "In quantum realism, the quantum wave is a processing wave and where a photon hits is the node reboot the server recognized," (which will be better understood upon reading Whitworth's *Quantum Realism* paper).
- "No matter how far apart the entangled photons are in space, they connect directly to their server source as pixels on a screen connect directly to the processing creating them."
- "Quantum realism requires the holographic principle. *The holographic principle doesn't mean the physical world is two-dimensional, as some assume, it means it is virtual. It is a consequence of how reality presents not how it operates. The physical world as an image must be delivered across two dimensions* [emphasis added]."
  And, "This is no Star Trek hologram one can enter and leave at will. Our bodies are its images, so if this hologram switched off, our physical bodies would disappear with it."

- "Matter is not solid. It is activity—processes and events—that is the essence of being." This is the essence of virtual reality as well.
- "We do not live in a world that always functions in a cause and effect way. The world consists of more than matter acting on matter." Quantum realism says that not only is there more than matter: it says that matter is to reality as pixels are to TV screens or icons are to computers. Appearances, symbols, but not having literal substance. Again, Schrödinger tells us, "What we observe as material bodies and forces are nothing but shapes and variations in the structure of space. Particles are just *schaumkommen* (appearances)." A fair descriptor of virtual reality.
- "There is always uncertainty, and the world is indeterminate. We live in a world of probabilities, not predictability." Equivalently, a virtual reality, emerging from quantum dynamics, would be uncertain, indeterminate, and manifesting out of (mathematical) probabilities rather than predictability.
- "Information and action can occur instantaneously across immense distances. It appears that an 'entangled' particle *instantaneously* 'knows' and reacts to the actions of its counterpart, even at unfathomable distances." Virtual reality is, as we have seen, information processing—non-physical actions and events, not dependent on physical laws and limitations. It is one entangled system: one mind.
- "Quantum events are uncaused, that is, self-caused. There is no cause outside of the quantum event itself that explains its action." So we opt for the miracle that the cosmos just appeared or the miracle that some intelligence created it. Quantum realism opts for the latter miracle. Quantum realism and virtual reality implies a "boot-up" that initiates reality, a boot-up of a virtual game, we might say.
- "We cannot predict events in advance of our observation." Virtual reality emerges out of quantum reality. It is a world of information processing and mathematical probabilities manifesting as they are consciously observed, no more amenable to prediction than quantum dynamics itself.

A Universe Full of Magical Things

- "What happens in the world is affected by our observation. The observer is not separate from the object being investigated... 'The so-called observer is actually a *participator*, an integral part of the quantum system.'" The observer processes information about mathematical probabilities and participates in creating the physical symbols that result. These symbols in the virtual "game" are the game's reality—our reality—but only allude to the underlying quantum, conscious reality. Our bodies are part of virtual reality; our consciousness, the observer, is not.
- "We only can say that something exists when we observe it: otherwise, as the philosopher Ludwig Wittgenstein has said, 'Whereof we cannot speak, thereof we should remain silent.'" In a virtual world, it is an observation that generates an appearance, as in the quantum world. (And in a virtual reality game, unless we observe it, it doesn't exist, no matter what we logically might want to conclude).
- "We subjectively participate in co-creating the universe. We each have our own unique consciousness. In quantum dynamics, there is no objectivity." In virtual reality, we have only our unique subjective experience, our unique subjective consciousness. Donald Hoffman, recall, has noted that "There are no public objects sitting out there in some preexisting space."
- "Observation is consciousness, and quantum systems *require* a conscious observer." The boot-up of our reality, our virtual "game," requires consciousness to initiate, and us to affirm.
- "The universe is interconnected at the quantum level—the micro world that underlies the macro world of classical mechanics—unfolding, complexifying, and evolving itself as it expands." A virtual world has no physicality, and thus no physical limits. There are no boundaries, there is only one reality, a quantum reality of expansive consciousness, mind, and, ultimately, only quantum mind. It is more than metaphysical, it is the core truth that physics, with reluctance, has concluded. It is worth noting how Whitworth understands the holographic principle Bohm spoke of: "Quantum realism requires the holographic

principle. The holographic principle doesn't mean the physical world is two-dimensional, as some assume, it means it is virtual. *It is a consequence of how reality presents not how it operates* [emphasis added]. The physical world as an image must be delivered across two dimensions" (Whitworth, 2014, *Quantum*). What is beyond the game is Bohm's implicate order, which we cannot easily access while playing the game. What we experience and relate to—and have access to in the game—is the explicate order.

– "The universe must somehow always have been alive, sentient, conscious, and self-reflectively aware. Because we have the capacity to know, obviously the capacity to know exists in the universe! These features occur as potential—superpositions, if you will—that emerge and evolve out of the booting up of our reality at the moment of the big bang. These features of consciousness likely precede, but are, at least co-existent with matter from the beginning." We are seeing that the virtual reality of quantum realism concludes that consciousness, quantum fluctuation, or Quantum Mind, is all that is, and that interpretations of quantum based in *an assumption of the primacy of matter* can never arrive at this knowing, because they fail to understand that *matter is only a symbol* of an underlying reality—not that reality itself. We humans are players, participant-observers in this virtual game, and are what is conscious, what is lived, and what is real. It is only our perceptions that are real.

BIOCENTRISM AND VIRTUAL REALITY

"… We have ignored a critical component of the cosmos, shunted it out of the way because we didn't know what to do with it. This component is consciousness."

—Robert Lanza.

I have referenced Robert Lanza and his book, *Biocentrism: How Life and Consciousness Are the Keys to Understanding the True Nature of the Universe,* quite frequently in the preceding pages. Even though he does not reference the "virtual reality" conjecture directly, it is implicit in all that he proposes in the concept of biocentrism. Lanza concludes his presentation on biocentrism with seven principles that embody his ideas, and which particularly align with and ratify the concepts of quantum realism, or virtual reality.

He elaborates on each:

"First Principle of Biocentrism: What we perceive as reality is a process that involves our consciousness. An 'external reality,' if it existed, would— by definition—have to exist in space. But this is meaningless, because space and time are not absolute realities but rather tools of the human and animal mind" (Lanza, 2009, p. 127).

He underscores here the virtual reality notion that the material world of space and time are subjective concepts, products of mind not truly existing anywhere (but in the game).

"Second Principle of Biocentrism: Our external and internal perceptions are inextricably intertwined. They are different sides of the same coin and cannot be divorced from one another" (Lanza, 2009, p. 127).

He here reiterates the virtual reality conclusion that the external world (the virtual game) is not concrete or real, but exists only in our minds. When we opt out of the game, it no longer exists—because it only has any reality or meaning in our minds.

"Third Principle of Biocentrism: The behavior of subatomic particles— indeed all particles and objects—is inextricably linked to the presence of an observer. Without the presence of a conscious observer, they at best exist in an undetermined state of probability waves" (Lanza, 2009, p. 127).

Here, Lanza calls on the essential conclusion of quantum mechanics which is that reality is determined by an observation from a conscious entity, and, further, that that observed reality has no existence outside of the observing entity, which aptly describes any virtual game as well.

"Fourth Principle of Biocentrism: Without consciousness, 'matter' dwells in an undetermined state of probability. Any universe that could have preceded consciousness only existed in the probability state" (Lanza, 2009, p. 127).

He is stating in this principle the idealist understanding of the primacy of consciousness, which is in line with the virtual reality conjecture, pointing to the illusory nature of matter. Quantum theory says conscious observation is required for what we experience to exist. Virtual reality as well illustrates that we can only perceive our perceptions, so that that is what we come to experience as real.

"Fifth Principle of Principal Biocentrism: The very structure of the universe is explainable only through biocentrism. The universe is fine-tuned for life, which makes perfect sense as life creates the universe, not the other way around. The 'universe' is simply the complete spatiotemporal logic of the self" (Lanza, 2009, p. 127).

The "player" in the virtual game is what gives the game existence and meaning. The game represents the identity of the player in the material-plane-fantasy, and as such gives life to the universe. Biological life is prerequisite for space-time and matter/energy and choice, all of which are real only in the game,

"Sixth Principle of Biocentrism: Time does not have a real existence outside of animal-sense perception. It is the process by which we perceive changes in the universe" (Lanza, 2009, p. 127).

The experience of time serves as a vehicle by which change is perceived in the holomovement that is the universe. Time highlights that all that is is now. Without an observer, the player, you and me, there is no time.

"Seventh Principle of Biocentrism: Space, like time, is not an object or thing. Space is another form of our animal understanding and does not have independent reality. We carry space and time around with us like turtles with shells. Thus, there is no absolute self-existing matrix in which physical events occur independent of life" (Lanza, 2009, p. 127).

Space and matter/energy are no more real without an observer than they are real in a virtual game. In a virtual game the idea of space is real only relative to the virtual objects in the game, and nothing more. Again, without the player or observer, without consciousness, the game ceases—our experience ceases.

The virtual conjecture casts a new light on "reality." In its assertion of the primacy of consciousness, virtual reality—quantum realism, biocentrism—gives rise to an understanding of our universe that is consistent, clarifying, grounding, and meaningful. We come to understand that what is real is a play within the mind, a process, and, as well, a deliberate and intentional "program" created around us in order for us to participate in authoring what the game, our reality, can become—as we will explore specifically in Chapters Seven, Eight, and Nine.

Virtual reality—that is, what our universe actually is—is a material-plane-virtual-game, wherein what we experience as physical, (the Materialist Paradigm), is only real within the game, wherein there is "really" nothing physical, objective, deterministic, or predictable. Our consciousness engages the (virtual) content of the game—and we, as players, interact as conscious, participatory observers, (the Consciousness Paradigm). This non-physical and virtual reality was birthed out of non-physical "quantum fluctuations in the infinitesimal vacuum," i.e., quantum mind, or consciousness itself, (The Quantum Mind Paradigm). All that we know, we know subjectively, and the location of all of our perceptions is in our consciousness, nowhere else. In this game, time and space and matter are "material" only for biological beings, (i.e. we who observe, which could go "all the way down)." Our conclusions about physicality and materiality are valid only in the game.

As we observed earlier, "What is real beyond the game is consciousness, in the 'form' of a creative 'programmer' who booted up this game... In the game all that is real, all that has any meaning, has to do with our individual points of consciousness, observing the game, and further, participating with the universe in what next evolves, as though we had free will."

SMOKE AND MIRRORS

So as we apprehend more deeply what quantum theory has unarguably described, we find that, quite logically, when we treat consciousness as primary and matter as emanating from consciousness, then quantum theory, in the guise of quantum realism, is completely consistent and provides us answers to the questions of science quite well, while offering a path toward unplumbed meaning and profound awe.

And in seeing consciousness as primary, we arrive as well at an inevitable conclusion: what we experience in our daily lives is unique to our own consciousness, and "who we are" is so different from and so much more than who we thought we were—a richer reality we are barely able to envision wherein what we experience as the physical plane is revealed to be only a virtual tease, a seductive representation of some greater ethereal truth that we, only now, can begin to explore.

If the universe is indeed a virtual reality, then there *is* another reality beyond our experience that is in charge of this simulation, even beyond Moses, Jesus, and Mohammed; beyond the Greek and Roman Empires; beyond the Catholic Church; beyond the billiard ball reality of the scientific revolution and the enlightenment; and even beyond an objective, material world. It is a smoke-and-mirrors world, wherein truth is beyond the images we witness—the things we see, smell, and touch; the interpretations we make; and the conclusions we draw. It is a Universe Full of Magical Things!

Our personal consciousness observes our (virtual) reality, these projections of mind. And our personal observations are collective, even with our nuanced subjective viewpoints: "One Mind is a collective, unitary domain of intelligence, of which all individual minds are a part," observed Dossey, as we noted.

Materialistic science, despite itself, has once more shown the flaws of its cherished "Standard Model," and, in its stead, Mind emerges— the Soul of our Universe: Lanza asserts that "…Experiments of the past decade truly seem to prove that Einstein's insistence on 'locality'—meaning that nothing can influence anything else at superluminal speeds—is wrong. Rather, the entities we observe are floating in a field—a field of mind… that is not limited by the external *space-time* Einstein theorized a century ago" (Lanza, 2009, pp. 52-53).

Walker (Walker 2000) eloquently argues—as do a host of others as we have repeatedly seen now—that this one mind is the quantum mind. The quantum mind, our consciousness, is at its very base what physicists are calling "quantum fluctuations." Quantum fluctuations are the actions and influences of consciousness itself. And we are all It.

When we accept that the objective world of physical matter is real, *but only to us who are in the game,* (wherein the objective physical world *is a* reality, but not *the* reality), we can consider that idea as a clue to a greater reality: a quantum world that has its own non-physical rules, keeping the structure of the physical game in check by non-material means, just as a programmer generates non-physical characters and structure in a virtual space and time.

Despite the digital, mathematical, rationalistic, and theoretical flavor we associate with "virtual," "simulated," "programmed," and other terms we've been using, in our reality we experience life, mind, self-reflective awareness, growth, change, love, relationship, heart, soul, spirit, contrast, suffering, and joy. *What is alive observes, chooses, and acts:* it is we, participant observers that are the universe, that make up the universe. It is our aliveness—our consciousness, our intelligence, our agency, and our participatory and creatively emergent processes, that make up what the living, dynamic, intelligent, conscious universe consists of!

To conceive, from a grand perspective, that our reality might be "created," or "programmed" by some greater consciousness does not compromise such motivators as values, meaning, or purpose. Or living, loving, learning, and laughing. It only clarifies. Simply, it permits us to give up the debate about first cause, and accept that we are somehow "other"-created, or perhaps, ultimately, self-created. It permits us to be clear that our existence here must be, by virtue of the limits of our knowledge and experience, a miracle. And thus it permits us to wonder "what meaning our being here might have?" as opposed to "does our being have a meaning at all?"

We will ahead describe a universe—our universe—that is a living being, intelligent and intentional, evolving and expanding in its innate wisdom and self-reflective awareness, deepening in its complexity. We are describing, if you will, *our* game, which emerges in all those ways. It seems to be "going somewhere," toward an increasingly compassionate manifesting of one mind as it persistently comprehends itself in richer and richer terms.

We have connected the dots and reasoned that we indeed do live in a virtual reality, with which we engage via our conscious participation. We consider this to be true from the vantage points of astrophysics and cosmology, physics, relativity, and quantum theory, and from the perspectives of metaphysics and consciousness studies as well. There is an inherent consistency we've seen among these disciplines, which is also ratified in the studies of biology, geology, mineralogy, evolutionary studies, developmental psychology, genetics, and other disciplines. In the next chapters we'll explore some of the terrain of these fields and describe how they too align with and confirm the living nature of our universe, and the essential consciousness of which it entirely consists.

Although this might feel to you like some strange new and unfamiliar and unreal world, dear reader, virtual reality reflects the conclusions and implications of quantum theory seamlessly. Within the limits of humanity's self-reflective capacity to make sense out of our world, quantum realism resonates in a way that makes Newtonian-inspired, materialistic science appear as if it were a broken-down horse-drawn wagon, hobbling to reach its destination. In contrast, our virtual game manifests as one consciousness, pervasive and dynamic. Rather than commit our energy and attention to defending an obsolete worldview, perhaps our efforts are better spent considering the living, conscious universe that is revealed within quantum realism, and contemplating the point of our being participants in such a game.

PART TWO:

THE POINT

"We are not our bodies, no! We are not our minds, oh!
How could we be the stories we tell, when words mean symbolically?
Beyond the mind, beyond describe, beyond understanding—a Higher Vibe—
We are the Souls that live within.
(Oh, the power Souls create, when they're free)."

—Julian Yeats, ("Higher Vibe," *Higher Vibe,* 2018).

Chapter Seven:

LIFE—THE ESSENTIAL
PROCESS OF THE COSMOS

LIFE EVOLVES AS THE ESSENTIAL PROCESS OF THE COSMOS

When Rolston describes what he calls the first big bang—that is, for him, the birth of matter—you can palpably feel his sense of astonishment at the miracle of something (so sensational!) seemingly from no thing. But his marvel pales in comparison to his wonderment about the birth of life, (his second big bang) and its origin and purpose: "What is novel on Earth is this explosive power to generate vital information. In this sense, biology radically transcends physics and chemistry...Such complexity involves emergence... a capacity and behavior of the whole that transcend(s) and (is) different from... the previous levels of organization" (Rolston, 2010, p. 45).

Truly, life is an emergence and a transcendence! But as we've established, it does not emerge from matter, but from the all-encompassing Quantum Mind. It is a phenomenon central to the universe, and yet science is confounded about its source, and even about what life actually is.[111]

---

111 We can describe life's attributes, as we will in the pages ahead, but we know nothing about its

"Growth and development are irregular and nonlinear;" says David Hawkins. "Practically nothing is known about the essential nature of growth, or any 'process' in nature for that matter. No one has ever studied the nature of *life itself*, only its images and consequences" (Hawkins, 2014, p. 237).

In fact, the (materialist) scientific definition of the concept "life" itself has been somewhat of a moving target—changing dependent on who is defining, and changing over time. Indeed, there is no commonly agreed on understanding in science of what, exactly, life is. An illustration, but not really a definition from Merriam-Webster, is: "the ability to grow, change, etc. that separates plants and animals from things like rocks or water."

The NASA website suggests that "some initial agreement is possible (among biologists, biochemists, geneticists, and cosmologists).

"Living things tend to be *complex and highly organized.*

"They have the ability to *take energy* from the environment and transform it for growth or reproduction.

"Organisms tend toward *homeostasis*: an equilibrium of parameters that define their internal environment.

"Living creatures *respond,* and their stimulation fosters a reaction-like motion, recoil, and in advanced forms, learning.

"Life is *reproductive*, as some kind of copying is needed for evolution to take hold through a population's mutation and natural selection.

"To grow and develop, living creatures need foremost to be *consumers* [emphases added], since growth includes changing biomass, creating new individuals, and the shedding of waste.

"To qualify as a living thing, a living creature must meet some variation for all these criteria."

It is noteworthy that in most general definitions of life, there is no mention of the requirement that a life form be carbon-based, which is the variety of life we are familiar with on Earth. So if we are looking to identify or define life, we see that we will want to be open to looking beyond only those forms we are familiar with.

"We won't come up with an adequate account of life until we have

---

essence. Note that Rolston defaults to the Materialist Paradigm which holds that life/mind (consciousness) evolved from matter. Our assertion in this book is the reverse: that matter evolved from consciousness, Quantum Mind.

examples of life as we don't know it. We have a single example of life, and you just can't generalize on the basis of a single example." said Carol Cleland, who studies the philosophy of science at the University of Colorado, Boulder (Moskowitz, December 10, 2010, "Life's great").

That may have just become possible. Chelsea Gohd, in the magazine *Futurism* in the December 2017 article "This Discovery Could Change the Way We Search for Alien Life," reported that "scientists have discovered a new bacterium that can survive solely off of chemicals in the air," and further says that "These organisms are so unique that they are opening up the potential for finding alien life, because... (it) could exist in much wider circumstances than previously thought" (Gohd, 2017, "This discovery").

Each of the criteria listed by NASA suggest action: creating complexity and organization, taking energy, seeking equilibrium, responding, reproducing, and consuming—each involving processes, events, activity, and all in the context of utilizing the surrounding environment. Life is indeed a verb, a process.

Further, life makes choices, as we'll see: it possesses agency. If that is so, all of life must possess some form of consciousness as well. Life grows, develops, and evolves, and it emerges. Mechanistic science is befuddled by life, which we have and will see, is conscious throughout. *What is alive observes, chooses, and acts.*

Evolutionary biologist Sahtouris, believes that life was always present in the universe: "Physicists now tell us...that the matter-energy of the earliest universe was already, by its very nature, bound to form living systems. Had things been just the tiniest bit different at the beginning, this would not be so, and we could not have evolved. *Perhaps then, life evolves as the essential process of the cosmos* [emphasis added] as a whole and is not just something happening at a special point we hunt for..." proposes Sahtouris (Sahtouris, 2000, p. 34). We will see that recent research has indeed detected organic molecules in the young universe.

Just as, in our endeavor here of trying to understand the cosmos which is our home, we started by looking at matter and found it to be illusory, now we are trying to get our arms around what life might be, and we find that it, too, is illusive. We might consider an interpretation for life that is far more powerful and potentially meaningful than that

A Universe Full of Magical Things

proposed by reductionist science—a comatose universe out of which a small oasis of lifeforms somehow, inexplicably, just shows up. Perhaps we might seriously consider the shocking, outrageous, and paradigm-shifting perspective that the universe *always* involved life, and that "life is the essential process of the cosmos."

Just as we have seen through the lens of physics that the universe is action, movement, and an evolving and essential process, we can see that life too involves exactly those features. *The vital creative movement of the emerging universe is itself more sensible and better understood if appreciated as a living process* as opposed to a lifeless mechanism. The universe is far more than merely a mechanical process.[112]

THE LIFE FORCE THAT IS PRESENT THROUGHOUT THE UNIVERSE

The notion of *the cosmos as a living being,* indeed a *unified* living being, is a perspective that has been adopted—perhaps surprisingly—in recent times by a wide variety of scientists from the various fields of computing, biology, cosmology, astrophysics, physics, epigenetics and others.

In a PBS documentary, *Through the Wormhole,* the MIT quantum computing engineer Seth Lloyd, who we have referred to previously, was interviewed by Morgan Freeman. (Freeman produced and hosted the program). Lloyd's research, as we've seen, has to do with "the interplay of information with complex systems, especially quantum systems," and he has performed seminal work in the fields of quantum computation and quantum communication, including proposing the first technologically feasible design for a quantum computer, demonstrating the viability of quantum analog computation. In his book, *Programming the Universe,* Lloyd contends that the universe itself is one big quantum computer producing what we see around us, and ourselves" (Freeman, 2010-2017, "Through").

Freeman says, "Lloyd developed and runs a quantum computer,

---

112 Understanding life as a *living process* verses a mechanical one, and one that pervades the universe, suggests that that living process, likely, transcends and includes carbon-based life we are so familiar with on Earth.

based on the fact that atoms and every subatomic particle can think. Seth argues that his computer proves that *subatomic particles can think,* and that the universe, which is entirely built from such particles must also be a quantum computer. It processes and stores information at the microscopic level on everything we see around us. And *if the universe is processing information, then it must be thinking. And it must be alive"* [emphases added] (Freeman, 2010-2017, "Through"). Lloyd goes further: "The universe is not only alive; it is more than alive. It contains life. It does all the things that living things do. It processes information. It moves energy from one place to another. Different pieces reproduce each other. …The universe as a whole can do much more than just what living things can do."

Morgan elaborates: "The human brain can perform to the power of 10^{16}, or 10 million billion computations in a second. In the same time, Seth believes, the universe performs about 10^{106} computations—which makes the universe impossibly smarter than we can ever imagine. Lloyd concludes: "If the universe is behaving like a great quantum computer—and let's face it, it is—then it's capable of any kind of complex behavior we can imagine: not just the creation of stars and planets, evolution of life. It's also capable of behaviors that we will probably never be able to comprehend."

"The universe," concludes Freeman, "could be the ultimate intelligent organism" (Freeman, 2010-2017, "Through").

We have pointed out that computation and processing are forms of thought, but "the action of using your mind to produce ideas, decisions, memories, etc." is a mode of thought beyond Lloyd's definition. As with materialist-oriented scientists in general, he does not recognize, factor in, or reckon with consciousness itself. It is quite a persuasive perspective to see the universe as "the ultimate thinking machine," but the reality of consciousness in the universe means that the universe is beyond being a data-computing machine. Nevertheless, the quantum computer notion of computations and processes Lloyd is working with involves consciousness, and it involves it beyond what science understands.

Interestingly, Lloyd clearly views the entirety of existence as one unified entity. Data processing and storage and computations are all programmed into a computer, which then performs functions that lead to outcomes—outcomes limited by the program. That seems to be sufficient for Lloyd to call such a computer a living and thinking machine.

Sahtouris sees life as so much more. She has suggested that we regard life as *a process that is self-creating* or *self-producing*, which then allows us to take a perspective that is different than the traditional materialist assumption that life emerged out of "dead" material. We can begin to see life from a more cosmic perspective, (Sahtouris, 2000, p. 34), a perspective that acknowledges life (and mind) as intrinsic and fundamental to the universe.

Elgin, too, is strongly of the point of view that the cosmos—everything that exists—is, indeed, a being that is living and unified. He notes, "Physicists once viewed our universe as composed of separate fragments. Today, however, despite its unimaginably vast size, the universe is increasingly regarded as a single functioning system," which to that extent aligns with Lloyd's point of view.

Elgin sees the unity in much the same way as Bohm does, observing that "Bohm sees animate and inanimate matter as inseparably interwoven with the life-force that is present throughout the universe, and that includes not only matter but also energy and seemingly empty space. For Bohm, then, even a rock has its unique form of aliveness. *Life is dynamically flowing through the fabric of the entire universe*" [emphasis added] (Elgin, 2000, Ions). Elgin portrays the universe as "…a unique kind of living system. I wouldn't say so much biological, but rather, *there's a deep aliveness that permeates and sustains the universe* [emphasis added]. And so we are more than biological beings, we are beings that also partake of the aliveness of the living universe," (Elgin, 2009, p. 215), echoing Lloyd's statement: "the universe as a whole can do much more than just what living things can do."

"Our enormous galaxy is a complex ecological system that nourishes star systems that in turn are producing planets that grow our forms of life," Elgin says. "Galaxies are cosmic gardens comprised of billions of stars in a complex, ongoing partnership with streaming gases and energy, all necessary to create the conditions for growing a rich diversity of life within the galaxy" (Elgin, 2009, p. 202).

Elgin is clear that "the universe (is) a living system that is being continuously regenerated and that has been *designed to support, in freedom, the evolutionary development of self-referencing and self-organizing systems at every scale* [emphasis added]. Instead of a deterministic universe," he says, "freedom is built into the quantum foundations of the universe" (Elgin, 2009, p. 202).

We find ourselves talking in terms of two different paradigms, one the Materialistic Paradigm, that sees the universe only as an inert physical realm, and the other, the Consciousness Paradigm, that sees a universe catalyzed by life and mind and action, movement, and flow, making choices in ever-emerging and new ways. This book is calling into questions the stark limits of the perspective and the validity of a materialist mindset. The Consciousness Paradigm clearly offers a richer, far more inclusive, expansive, and abstract assessment of our universe and what it and we are capable of creating.

The capacity to make choices, to have agency, we will see, is present at every developmental level within the universe. Healthy living systems maintain a vital dynamic that goes beyond the models of chemistry and physics. Healthy living systems utilize information from their environment in proactive, agentic ways, through DNA, according to Rolston: "What makes the critical difference is not the matter or the energy, necessary though they are; what makes the critical difference is the information breakthrough with a resulting capacity for agency, for doing something." (Rolston, 2010, pp. 43-44).

As if intended to bring this point to even greater clarity, Elisabet Sahtouris in *EarthDance* states that "Jeremy Narby, an anthropologist and writer of several books who spent several years living in the Amazon rainforest,…calculated…that a single handful of living earth contains more DNA [that is, bio-information] than that of our entire bodies—because bacteria are packed far more closely in soil than cellular nuclei are packed in our bodies" (Sahtouris, 2000, p. 377).[113]

We need to consider the concept of life more broadly and deeply. We need to look at alternative paradigms that have the capacity

---

113 Laura Pritchett, in an essay entitled, "Back to the Basics: An Ode to Germs, Guts, and Gardens," in *High Country News,* Nov. 9, 2015,Vol. 47, No.19, writes, "…soil fertility and human health are directly related; the word 'human,' after all, is related to 'humus,' meaning 'earth or soil.' Humans are just beginning to understand what that means—both the human gut and the layer of dirt are somewhat unexplored territory. There are 100 trillion bacterial cells in a human body, 10 times the number of human cells—on a DNA level, at least, we are more microbial than human. There are 100 million to 1 billion bacterial cells in one teaspoon of healthy soil…" (Pritchett, 2015, "Back to").

to expand our understanding of our universe and of existence itself. "The most promising definition of life among biologists..." affirms Sahtouris, in *EarthDance*, "seems very nearly to fit galaxies, if not stars. This definition we owe to the Chilean biologists Humberto Maturana and Francisco Varela. Their concept of life is a process called autopoiesis, (pronounced auto-po-EE-sis), which in Greek means self-creation or self-production. An autopoietic unity...produces the very parts of which it is made and keeps them in working order by constant renewal" (Sahtouris, 2000, p. 33).

In the 1970s with the advances in systems science the idea was suggested that order appears spontaneously in the universe because of the tendency of natural systems to self-organize. But the process by which the new information that would lead to such self-organizing was still unaddressed, which led Rolston to note: "What is inadequately recognized in the 'self-organizing' accounts is that, though no new matter or energy is needed for such spontaneous organization, new information is needed in enormous amounts and that one cannot let this information float in from nowhere. Over evolutionary history, something else is going on 'over the heads' of any and all of the local, individual organisms. More comes from less, again and again." Rolston believes "a more plausible explanation is that, complementing the self-organizing, there is a Ground of Information or an Ambiance of Information, otherwise known as God" (Rolston, 1999, p. 355). Bohm's Holographic Universe model would point to the Implicate Order as the source of new information. The zero-point field described by McTaggart suggests the source of this ongoing informational download. Additionally, the interplay among each of the living systems that participate in creating a given ecosystem itself also offers a rich, complex pool of information that might be used by self-producing and self-creating entities in their agentic and emerging journeys.

LIFE'S BEGINNINGS ON EARTH

Far and wide, those who have tried to comprehend the origins of life have come away overwhelmed. The famous molecular biologist, biophysicist, and neuroscientist Francis Crick, who co-discovered of the structure of the DNA molecule in 1953, remarked, incredulous: "In some sense, the origin of life seems almost a miracle, so many are the conditions that would need to be satisfied to get it going" (Watson & Crick, 1953, "A structure").[114]

Because of the conditions, timing, and complexity required for life to somehow emerge, many have concluded that we likely will never irrefutably know where and how life originated. "Unfortunately, it is impossible to obtain direct proof of any particular theory of the origin of life," according to James D. Watson, the other co-discoverer of the structure of DNA.[115] "The sobering truth is that even if every expert in the field of molecular evolution were to agree how life originated, the theory would still be a best guess rather than a fact" (Watson, 1987, p. 116).[116]

---

114 Crick, in 1962, was (jointly) awarded the Nobel Prize for Physiology or Medicine for his work with the molecular structure of nucleic acids.

115 Watson shared the above award with Crick.

116 And what do we know about the formation of DNA and RNA? We are still learning. DNA, or deoxyribonucleic acid, provides the biological guidelines for any functional living organism. RNA, or ribonucleic acid, facilitates the process. RNA is more versatile than DNA, capable of performing a wide variety of tasks in an organism, but DNA is more stable and stores more complex information for longer periods of time. We will describe ahead in a variety of contexts the significance and the mystery of DNA in life as we know it.

Writing in *The New Scientist*, Michael Marshall says, "Chemists are close to demonstrating that the building blocks of DNA can form spontaneously from chemicals thought to be present on the primordial Earth. If they succeed, their work would suggest that DNA could have predated the birth of life. DNA is essential to almost all life on Earth, yet most biologists think that life began with RNA. Just like DNA, it stores genetic information. What's more, RNA can fold into complex shapes that can clamp onto other molecules and speed up chemical reactions, just like a protein, and it is structurally simpler than DNA, so might be easier to make. After decades of trying, in 2009 researchers finally managed to generate RNA using chemicals that probably existed on the early Earth. Matthew Powner, now at University College London, and his colleagues synthesised two of the four nucleotides that make up RNA. Their achievement suggested that RNA may have formed spontaneously – powerful support for the idea that life began in an RNA world.

"Powner's latest work suggests that a rethink might be in order. ...He is trying to make DNA nucleotides through similar methods to those he used to make RNA nucleotides in 2009. And he's getting closer. Prebiotic chemists have so far largely ignored DNA, because its complexity suggests it cannot possibly form spontaneously...Conventional wisdom is that RNA-based life eventually switched to DNA because DNA is better at storing information. In other words, RNA organisms made the first DNA.

"If that is true, how did life make the switch? Modern organisms can convert RNA nucleotides into DNA nucleotides, but only using special enzymes that are costly to produce in terms of energy and materials. 'You have to know that DNA does something good for you before you invent something like that,' Switzer says. He says the story makes more sense if DNA nucleotides were naturally present in the environment. Organisms could have taken up and used them, later developing the tools to make their own DNA once it became clear how advantageous the molecule was – and once natural supplies began to run low. 'Early organisms must have scavenged for materials in this way,' says Matthew Levy of the Albert Einstein College of Medicine in New York City. 'The early Earth

Steve McIntosh, in his brilliant synthesis, *Evolution's Purpose: An Integral Interpretation of the Scientific Story of Our Origins,* with regard to DNA, makes this significantly consequential point about our current perspective on evolution:

"When we examine certain forms of evolutionary organization that have not arisen through the mechanisms of natural selection, purely physicalist explanations are…inadequate. The classic example of the unexplained influence of information and the role of formative causation in prebiotic evolution is found in the case of DNA, which emerges prior to the first reproducing living cells. The DNA molecule has the *highest information density* of any known form of matter. Nothing like it is found anywhere else in prebiotic physics or chemistry. Each one of the 'linguistic molecules' is saturated with information in an amount roughly equivalent to a large set of encyclopedias. And not only does the DNA molecule store massive amounts of data, it also includes the code which governs the retrieval of the information it carries. How this amazing form of molecular organization evolved *without the mechanisms of biological natural selection* remains a profound mystery" [emphasis added] (McIntosh, 2012, p. 48).

Rolston, again, tells us that: "This biological sense of information is proactive, agentive…What makes the critical difference is the information breakthrough, with resulting capacity for agency, for *doing* something" (Rolston, 2010, pp. 43-44). Darwinian and Neo-Darwinian versions of evolution with natural selection and random mutation being the only influencing dynamics in its theory—hypothesis really—don't speak to or consider information and agency and emergence at all. With such a doctrine, a rich understanding of life and mind is not discernable.

The conundrum of the essence and origins of DNA will likely lead us eventually in the direction of the zero-point field and the quantum

---

was probably a bloody mess,' he says, with all manner of rich pickings on offer.

   "Powner suggests another alternative. Life may have begun with an 'RNA and DNA world', in which the two types of nucleotides were intermingled. Powner's co-author Jack Szostak, of the Harvard Medical School, has shown that 'mongrel' molecules containing a mix of DNA and RNA nucleotides can perform some of the functions of pure RNA…Powner suggests that life started out using these hybrid molecules, gradually purifying them into DNA and RNA. Right now, though, there's nothing to tell us exactly how and when life first used DNA. 'It almost becomes a choose-your-own-adventure game,' says Levy" (Marshall, August 22, 2012, "DNA could").

fluctuations that underlie our reality, a (quantum) dynamic that will likely address, as well, in time, the many inadequacies and discrepancies of our current neo-Darwinian spin on evolution.

Sahtouris persuasively argues that "Darwinism—and its updated version Neo-Darwinism—is a misleading way of seeing nature. The notion of the separateness of each creature, competing with others in its struggle to survive, had well described and justified English and American societies' new forms of competitive and exploitive industrial production in a world of scarcity. But we are now beginning to understand that humans must learn to harmonize our ways with those of the rest of nature instead of exploiting it and each other ruthlessly... The Darwinian theory of evolution was applied to forming a society, a social system, designed in accord with, and justified by the Darwinian concept of nature. If we learn to see evolution as a single holarchy of holons working out of the mutual consistency of cooperative health and opportunity, we can set up a social system to match that view" (Sahtouris, 2000, pp. 108-109).

Further, and more to our point, as long as evolution continues to be understood using the materialist, reductionist, deterministic, paradigmatic assumption of matter first, it is more than unlikely the riddle of DNA can be solved. And as we are seeing and will see, should we open ourselves the notion of the primacy of mind, it becomes easy to visualize how the intelligence that pervades nature participates and cooperates in creating and generating an informed *emergence,* just as an electron manifests as particle or wave with our participatory observation. *Mind—consciousness—creates, deliberately and intelligently.*

To understand the essence and origins of DNA may involve the same sorts of questions as understanding the essence and origins of life. That understanding may, in the end, be beyond our capacities. Lloyd, for example, notes specifically some of the major obstacles to our *ever* understanding the origins of life, from his point of view: "You would not expect...to reproduce the exact origins of life because

(A) there are many possible sets of initial conditions,
(B) the set of reactions could be driven in many different ways,
(C) we don't know what these conditions are,

(D) there's a huge number of possible ones,

(E) because these interactions are non-linear and hence

(F) chaotic in lots of cases, so that

(G) they can be very sensitive to these initial conditions" (Lloyd, August 27, 2007, "Life: what").

Life's origins can seem to be inscrutable. "How remarkable is life?" asks the Harvard chemist, George Whitesides. "The answer is: very. Those of us who deal in networks of chemical reactions know nothing like it…How could it be that any cell, even one simpler than the simplest that we know, emerged from the tangle of accidental reactions occurring in the molecular sludge that covered the prebiotic earth?" (Barrow, *etal.*, 2008, pp. xi-xix).

Further, Christian de Duve notes, "If you equate the probability of the birth of a bacterial cell to that of the chance assembly of its component atoms, even eternity will not suffice to produce one for you" (de Duve, 1984, p. 356).

And yet, Nick Lane[117] hypothesizes that the beginnings of life *on Earth* might be knowable after all, and may have arisen, in the right conditions, out of inert, inorganic matter. He says that recent evidence from various fields, including biochemistry, molecular biology, and geology, suggests that energy is the key ingredient that might generate the whole process. In his dense, but clear book, *The Vital Question: Energy, Evolution, and the Origins of Complex Life*, (Lane, 2015). he first acknowledges that there are countless definitions of life over which science may quarrel, (among which is that of autopoiesis), then puts forward the idea that energy is the overlooked ingredient that is requisite in any worthy notion of what we might mean by "life." Energy underlies the intake of nutrients and expulsion of waste which constitute essential functions of life, and it involves the metabolism of chemistry, the capacity to maintain life by reproduction, along with the establishment of the physical nest within which life might come to be, among other features we have discussed. Indeed, he says, it is energy that is indispensable for the generation of the vital information that is the

---

117 Nick Lane is a chemist, lecturer, and honorary researcher in evolutionary biochemistry at University College London. He has published four books which have won several awards.

currency of that which lives.[118]

Lane sees life beginning with the creation of a sustainable living cell necessitating a containing membrane, (like a skin); it also needs to be able to collect, access, and use genetic instructions; and it must establish a way of taking energy from the environment and putting it to work to run the cell's processes. Lane notes that it is energy itself that allows for each of these things to develop.

We have come to understand the specifics of biochemical ingredients essential to the establishment of life, Lane says. And these biochemical ingredients are intimately bound up in the minerals and chemicals preexistent in the environment, and which, as well, after interacting with primitive life forms, resulted in still new kinds of minerals. Once primitive life forms took hold in the Earth's environment, the environment complexified, powering the forces of evolution.

The technology of x-ray crystallography has permitted the charting of large proteins and energy reactions in minute detail, leading to more precise questions about the protein complexes, mineral structures, enzymes, and so on required for a viable living cell, says Lane. These are necessary (though not sufficient) ingredients for the development of life. He points to environmental conditions *and* an energic process that might serve as the womb of life.

"The earth was initially very hot, and without an atmosphere," Stephen Hawking writes, setting the stage: "In the course of time, it cooled and acquired an atmosphere from the emissions of gases from the rocks. This early atmosphere is not one in which we could have survived. It contained no oxygen, but it did contain a lot of other gases that are poisonous to us… Primitive forms of life…flourish[ed] under these conditions. It is thought that they developed in the oceans, possibly as a result of chance combinations of atoms into large structures, called macromolecules" (Hawking, 1988, p. 120).

Lane, concurring, states that the most likely location for where life

---

118 Etiologically, energy and matter are obviously necessary for the beginnings of life, but both Lane and Rolston iterate that the truly distinctive feature in the calling forth of life is, as Lane declares, "the generation of vital information." But Rolston's perspective highlights the agency that is generated by information: "What makes the critical difference is not the matter or the energy, necessary though they are; *what makes the critical difference is the information breakthrough with resulting capacity for agency*, [emphasis added], for doing something" (Rolston, 2006, Introduction). Any adequate definition of life must incorporate a capacity for agency: without agency there is no life.

might have begun on Earth were the alkaline hydrothermal vents on the ocean floor in the vicinity of the Mid-Atlantic Range, along with a few other such formations. These, which were discovered recently in 2000, have the right combinations, he believes, of warm energetic minerals pouring out of the ocean bed and forming calcium carbonate chimneys full of micropores, and comprising the constituents required to create organic chemicals, the precursors of life.

Mineralogist Bob Hazen describes how he believes that process likely has unfolded in the *Smithsonian* article, "The Origins of Life: A mineralogist believes he's discovered how life's early building blocks connected four billion years ago." The author, Helen Fields, writes: "There were only about a dozen different minerals—including diamonds and graphite—in dust grains that pre-date the solar system, Another 50 or so formed as the sun ignited. On Earth, volcanoes emitted basalt, and plate tectonics made ores of copper, lead and zinc.

"'The minerals become players in this sort of epic story of exploding stars and planetary formation and the triggering of plate tectonics,' Hazen says. 'And then life plays a key role.' By introducing oxygen into the atmosphere, photosynthesis made possible new kinds of minerals—turquoise, azurite and malachite, for example. Mosses and algae climbed onto land, breaking down rock and making clay, which made bigger plants possible, which made deeper soil, and so on. Today there are about 4,400 known minerals—more than two-thirds of which came into being only because of the way life changed the planet. Some of them were created exclusively by living organisms… 'The origin of life uses rocks, it uses water, it uses atmosphere. Once life gets a foothold, the fact that the environment is so variable is what drives evolution'" (Fields, October, 2010, "The origins").

On Earth life emerged—and is co-created—out of the complex interactions among all of the features of the ever morphing environment, including the importance of energy, as Lane cites, and which Sahtouris describes as: "the very rich organic chemistry present in the young solar system," and throughout the universe. "Our planet never was a ready-made home, or habitat," she explains, "in which living creatures developed and to which they adapted themselves. For not only does rock rearrange itself into living creatures and back, but living creatures also arrange rock into habitats—into places comfortable enough for them to live and multiply" (Sahtouris, 2000, p. 37).

"To make things more complicated, much of the rock that is transformed into live creatures is later transformed back into rock. And so, just as creatures are made of atoms that were once part of rock, almost all rocks on the Earth's surface are made of atoms that were once part of creatures," she says (Sahtouris, 2000, p. 36).

Current thinking is that the process got started, both Lane and Hazen argue, deep in the ocean. "'We've got a prebiotic ocean and down in the ocean floor, you've got rocks,' visualizes Hazen. 'And basically there's molecules here that are floating around in solution, but it's a very dilute soup.' For a newly formed amino acid in the early ocean, it must have been a lonely life indeed. The familiar phrase 'primordial soup' sounds rich and thick, but it was no beef stew. It was probably just a few molecules here and there in a vast ocean.

"'So the chances of a molecule over here bumping into this one, and then actually a chemical reaction going on to form some kind of larger structure, is just infinitesimally small,' Hazen continues. He thinks that rocks—whether the ore deposits that pile up around hydrothermal vents or those that line a tide pool on the surface—may have been the matchmakers that helped lonely amino acids find each other, [spawned by the energic movement of the moon, seas, tides, and plates, geothermal and hydrothermal and a variety of other actions]. Rocks have texture, whether shiny and smooth or craggy and rough. Molecules on the surface of minerals have texture, too. Hydrogen atoms wander on and off a mineral's surface, while electrons react with various molecules in the vicinity. An amino acid that drifts near a mineral could be attracted to its surface. Bits of amino acids might form a bond; form enough bonds and you've got a protein" (Fields, October, 2010, "The origins").

Lane's research and Hazen's conclusions might be seen as verification of a mechanistic universe, in that they describe life potentially emerging from an inorganic base. But, however the first manifestations of what we consider "life" unfolded, they unfolded because of the inherent potential of the universe to produce life, as Hazen submits. Lane's notable ingredient, energy, is undeniably essential for the earthly processes and interactions that have resulted in life. But what is the source of this energy? Quantum physics would point to the quantum fluctuations that underlie our universe, a universe of movement and action and events, yielding the information and information processing which Lloyd has described as thinking and alive throughout, as an essence, not merely an attribute.

Because of the complex quantum interaction of all aspects of the environment of Earth as a whole, life emerged on the planet—out of the increasingly integrated and complexifying system that is Earth itself. Within this environment, "The basic molecules of life, it turns out, are able to form in all sorts of places: near hydrothermal vents, volcanoes, even on meteorites. Cracking open space rocks, astrobiologists have discovered amino acids, compounds similar to sugars and fatty acids, and nucleobases found in RNA and DNA. So it's even possible that some of the first building blocks of life on earth came from outer space." maintains Hazen, (prophetically, it turns out), as a result of his "investigating how the first organic chemicals—the kind found in living things—formed and then found each other nearly four billion years ago [on Earth]" (Fields, October, 2000, "The origins").

Indeed, Hazen's speculation about the first building blocks of life on Earth perhaps coming from outer space has been the subject of intense study. Now, in fact, paradigm altering and powerful recent research from the ALMA radio telescope suggests it may be time to alter our present thinking about *where* in the universe life may have originated. As Nola Taylor Redd recently reported on the space.com website, "For the first time, astronomers have discovered complex organic molecules, the basic building blocks for life, in a disk of gas and dust surrounding an alien star...The giant ALMA radio telescope has detected complex organic molecules—the building blocks of life—suggesting that the conditions necessary for life are universal...The prolific amount of material reveals that Earth's solar system is not the only one to contain these complex molecules, suggesting that *the ingredients required for life to evolve may exist throughout the universe*" [emphasis added] (Redd, April 8, 2015, "Basic ingredients").[119]

---

119 Nola Taylor Redd, Space.com contributor, reports "For the first time, astronomers have discovered complex organic molecules, the basic building blocks for life, in a disk of gas and dust surrounding an alien star. To the researchers' surprise, the organics found around a young star called MWC 480

Karin Öberg, lead author of research at Harvard-Smithsonian Center for Astrophysics in Massachusetts, elaborates: "The very rich organic chemistry present in the young solar system…is far from unique. It thus seems likely that the prebiotic chemistry that took place in the solar system, including Earth, is also happening elsewhere" (Redd, April 8, 2015, "Basic ingredients").

So recent data lead to a conclusion that essential catalysts for the beginnings of life, in the form of complex organic molecules, pervade and pervaded the universe, leading, over time to the evolution of life in myriad forms—not suddenly, as mechanistic science tells the story: "Pop! Now life!" The emergence of life forms on Earth, and of life forms of ever-increasing complexity, we are finding, involved a process of the sowing of organic molecules throughout the early universe, eventually seeding Earth, along with a complex variety of material and temporal, energic and environmental, quantum and informational, and myriad other features, together orchestrating a unified living process on our Earth.

COMPLEXITY AND EMERGENCE OCCUR HOLISTICALLY, AS ONE INTEGRATED LIVING SYSTEM

It appears, then, that the universe as a whole has accommodated the potential for life—not merely here or there, but throughout. And Earth, too, has accommodated the potential for life—throughout. Life is a process that happens everywhere on our planet: "Planetary life is not something that happens here or there on a planet—it happens to the planet as a whole," declares Sahtouris (Sahtouris, 2000, p. 39). Earth itself is one integrated living system.

Tom Chi of Google Glasses illustrates this notion descriptively:

---

are not only surviving but thriving in quantities slightly higher than those thought to have existed in the early solar system. The prolific amount of material reveals that Earth's solar system is not the only one to contain these complex molecules, suggesting that the ingredients required for life to evolve may exist throughout the universe… After planets in the early solar system evolved, they suffered constant bombardment from icy comets. These comets, leftovers from the birth of the solar system, are thought to have seeded Earth with water and other elements necessary for life to evolve. By studying the composition of comets today, scientists can determine how plentiful complex organic molecules, such as methyl cyanide, were in the early solar system" (Redd, April 8, 2015, "Basic ingredients").

A Universe Full of Magical Things

"Humans lose and replace a little more than a kilogram a day of solid material, much of that through the process of breathing. And the air we breathe reaches halfway across Earth in four to five days. Some of the air we are breathing now was in the air on the other side of the planet four or five days ago. Some of the air we breathe today will make up atoms of a flower growing on the other side of the earth in four or five days: we are and it is all connected. It is and we are all one" (Truthinsideofyou.org).

Lipton and Bhaerman, in their book *Spontaneous Evolution,* observe as well that "Earth and the biosphere—and that includes us—are an integrated living system" (Lipton & Bhaerman, 2009, p. xii).

A healthy living system, such as Earth is, shows dynamic vitality on a large scale despite components being perceived as "non-living" says Sahtouris (Sahtouris, 2000, p. 39). She goes into further detail: "...plants and animals contain much supporting 'nonliving' matter in their woody trunks and shells and bones, their thorns and hoofs and nails, their hair and scales. Nonliving planets may also be very much a part of live galaxies, perhaps even play a structural role in their dynamics...What about the earth itself? Many scientists argue that it cannot be a living being because only its outermost layer...shows signs of life. What then, we may ask, about a redwood tree, which is ninety-nine percent dead wood with just a thin skin of life on its surface? No one argues that redwoods are not alive. It is new in modern science to look at the cosmos and the nature of our planet in this way" (Sahtouris, 2000, pp. 35-36).

Everything existing in a given environment is part of and fundamentally contributes to the life forms that are present. Sahtouris points out that "every molecule of air you breathe...has been recently produced inside the cells of other living creatures! Thus the atmosphere is almost entirely the result of the constant production of gases by [living] organisms. If they stopped...the atmosphere would burn itself out rather quickly" (Sahtouris, *Beyond Darwin, Part 1*).

The National Geographic Channel series, *The Known Universe,* adds: "The atoms that form the planet have been reused to create and sustain every organism that ever lived. When you take a glass of water and you drink those oxygen atoms, really what you are drinking is the remnant of star explosion. Of course, since those oxygen atoms came to Earth, they've been recycled a good deal too, so you're also drinking bits of dinosaurs and bits of all the famous people you've ever heard of.

"Each time you take a breath, quadrillions of atoms enter your lungs. These atoms have been mixed throughout the Earth's atmosphere over time, but they're still the same nitrogen and oxygen atoms that were there... Each one of the trillions upon trillions of atoms inside us began their journey inside a star. Since then, they've made countless stops in mountains and rivers, in trees and blades of grass, in molds and mushrooms, in deer, daffodils, and dinosaurs. And when the atoms are done in us they'll go on to recombine with something else. That's how it's been since the beginning, a continual process of creation and re-creation (National Geographic Channel, 2009, "The known").

"Every atom that is now, ever was, or ever will be a part of us will live on somewhere in Gaia's ever evolving dance for billions of Earth years yet to come. Even after the Earth dies, those atoms will live on as part of our galactic dance, some perhaps finding their way into new living bodies of new planets," declares Sahtouris (Sahtouris, 2000, p. 135). That creation and re-creation process involves everything material in the entire universe—yet another vivid documentation that everything is interconnected!

Sahtouris writes, "On the Earth's surface, scientists have a hard time finding any rock that has not been part of living organisms, that was not transformed into living matter before it became rock again... Thus, the molecules in virtually all of the atmosphere, all of the soils and seas, all of the surface rocks and much of the underlying recycling magma, have been through *at least* one phase in which they were living creatures!" (Sahtouris, 2000, pp. 80-81).[120] Evolution, increasing com-

---

120 A recent article entitled "195-million-year-old dinosaur bone yields traces of soft tissue," notes that "Scientists have discovered fragments of protein inside a 195-million-year-old dinosaur bone, by far the oldest such traces of soft tissue ever detected... The significance of the find is that the team was able to find any sign at all of organic material in such an old specimen" (Semeniul, Jan. 31, 2017, "195 million").

This discovery demonstrates that great numbers of "rocks" on the surface of the Earth consist of fossils, which are, according to Wikipedia, " the preserved remains or traces of animals, plants, and other organisms from the remote past...and their placement in *fossiliferous* (fossil-containing) rock formations and sedimentary layers (strata) is known as the *fossil record*." So when Sahtouris argues that "rock rearrange(s) itself into living creatures," the intertwining of organic and inorganic materials which arrange themselves into life at one point in time (or not, at another), cannot truly be

plexification, and emergence unfold persistently through time.

Everywhere he looks, Hazen says, he sees this fascinating process of increasing complexity. 'You see the same phenomena over and over, in languages and in material culture—in life itself. Stuff gets more complicated'" (Fields, October, 2010, "The origins").

So where, then, is the line that separates living from non-living? Freeman Dyson offers the point of view that "the nonliving universe is as diverse and as dynamic as the living universe" (Dyson, 2015, p. 7). Over the course of time, every atom, molecule, and object is intimately associated with all that we come to call "life."

Instead of a worldview in which inanimate rocks unaccountably morphed into purposeless intelligence and awareness, Sahtouris argues for a rich, thriving, and integrated worldview of the universe, as alive throughout; a universe out of which life, along with consciousness and the immense brilliance of human mind, has emerged.

Elgin says that "what we're discovering through both the lens of science and the lens of spirituality is that the universe is not a nonliving machine: *it is a living subjective presence* [emphasis added] and what we are doing is learning how to live in a living system" (Elgin, May 9, 2009, "The living").

In his books, Elgin also argues for an expanded description of life, a holistic paradigm, for both Earth and the universe that includes features such as these:

- It is unified,
- There is energy flowing throughout,
- It is continuously regenerated,
- There is sentience (consciousness) throughout, and
- It is able to reproduce itself.[121]

---

separated from each other. They are only separate aspects of a singular process, which is the oneness of all of life on Earth.

121 The idea that the universe might reproduce itself is based in current research and mathematics suggesting that black holes may play a key part in the life cycle of galaxies and even our universe. Specifically, Elgin speculates that the phenomena of black holes may be inverse big bangs (capable of generating "daughter universes)" and are the likely cause of our big bang. Astrophysicists have learned there is a massive black hole at the center of our galaxy, dubbed "Sagittarius A," which may be a crucial aspect in the life cycle of our galaxy.

    With regard to the same question applied to Earth, Sahtouris has this to say, (Sahtouris, 2000, p. 82): "we have come to think of growth and reproduction as essential features of living beings. The

Further, Elgin believes that our cultural understanding of a dead universe is both obsolete and serves as a poor starting point for understanding our universe and our existence. So when we speak of *a living universe*, we are saying it *cannot* be dead, inert, or chemical.

Kauffman, in *Reinventing the Sacred,* says it beautifully as well: "Life, and with it agency, came naturally to exist in the universe. With agency came values, meaning, and doing, all of which are as real in the universe as particles in motion. 'Real' here has a particular meaning: while life, agency, value, and doing presumably have physical explanations in any specific organism, *the evolutionary emergence of these cannot be derived from or reduced to physics alone.* Thus life, agency, value and doing are real in the universe. This stance is called emergence.... A couple in love walking along the banks of the Seine are, in real fact, a couple in love walking along the banks of the Seine, not mere particles in motion" (Kauffman, 2008, p. x).[122]

Sahtouris considers the following principles to be inherent in life and in all healthy *living systems*:

- Autopoiesis (or self-creation),[123]
- Related to parts as a whole and as part of other wholes (holarchy),
- Self-knowledge,
- The capacity to self-regulate or self-maintain,
- Ability to respond to stress,
- Transformations of matter, energy, and information,
- Communication among parts,
- Empowered use of all parts,

---

autopoietic definition of life, remember, does not include them as essential or defining features; rather, they are consequences of the autopoietic life process—something that may or may not happen, as when people do or do not reproduce. Therefore, the argument that the Earth cannot be alive because it does not grow or reproduce does not hold." This argument applies as well to the universe itself.

László's concept of the metaverse references potential processes of growth and reproduction, whereby we are the results of countless parent universes that have preceded us.

122 Building on our ongoing theme of the inadequacy of the Materialist Paradigm, author and biochemist Michael Denton points out the contrast between what is possible in the natural world verses what is possible in human technology, saying that, "Every living system replicates itself, yet no machine possesses this capacity to the slightest degree." (Kurzweil & Ray, 2001, p. 84).

123 Once more, autopoiesis describes "self-creation" or "self-production," an internal process, as distinct from "reproduction," involving external offspring. "An autopoietic unity produces the very parts of which it is made and keeps them in working order by constant renewal." (Sahtouris, 2000, p. 33). Sahtouris notes further that "autopoiesis as a definition of life does not include growth or reproduction, though these are features of many living entities. One can be alive without reproducing...." (Sahtouris, 2000, p. 72).

- Coordination among parts,
- Negotiated self-interest at all levels of hierarchy,
- Reciprocity and mutual assistance of parts,
- Ability to conserve what works well, and
- The capacity to change creatively (Sahtouris, 2000, p. 369).

This is a fairly broad list of descriptors of characteristics of living systems—living beings—like tomato plants, bacteria, humans, insects, dolphins, planets, and galaxies, for example. According to criteria such as these, it makes sense to understand our home environment as a living entity: matter, energy, information—and growing and knowing—in order to "do something," that is, having agency.

Life is complex, has agency, (and so, consciousness), and is emergent and evolving. It is an inclusive, unitary process, as, too, is the universe itself—and for us, particularly, on Earth. For Sahtouris, "our planet and its creatures constitute a single, self-regulating system that is in fact a great living being" (Sahtouris, 2000, p. xvii).

In the next chapters, we'll examine how this singular being moves through evolution—differently than the prevailing materialist worldview—toward an ever-unfolding, synchronous emergence. We will find that the Materialist Paradigm we've been taught to assume is true cannot contain the observations of a living, choosing, unified, and aware interconnected process. We'll see that a Consciousness Paradigm better reflects the results gleaned from recent scientific research in a variety of disciplines, and provides us with a congruent, sensible, and exciting lens for making sense of what the universe is actually showing us.

Chapter Eight:

—◆◆◆—

THE CAPACITY TO KNOW—
INTELLIGENCE IN NATURE

EARTHLIFE AND THE CAPACITY TO KNOW

"If the world is composed of matter that feels—of stuff that tingles with the spark of spirit all the way down to cells, molecules, atoms, and subatomic particles—then the world around us teams with sentient beings, themselves full of intrinsic meaning, constantly sending out messages as a kind of 'connective tissue' that unites all of nature in a universal network of shared, participatory meaning."

—Christian de Quincey (de Quincey, 2005, p. 91).

"All living things—including even those entirely devoid of nervous systems—can (and must) use some form of intelligence to survive. In fact, I believe that intelligence and the living process are one and the same: to live, organisms (or communities of organisms,) must absorb information, store it, process it, and develop future strategies based on it. In other words, to be alive, one must think."

—Neurosurgeon Frank Vertosick, *The Genius Within* (Vertosick, 2002, p. 4).

"Intelligence and the living process are one and the same."

In *The Genius of Birds*, Jennifer Ackerman states that life itself involves information processing, and differentiates between "innate hard-wired behavior" (calculation, instinct, programming) and "learned innovative behavior," (thinking, adaptation, innovation), and pointing out that these are properties that exist in *all* living beings (Ackerman, 2016, p. 52).[124] We can see that the natural world involves an elaborate and on-going dynamic of processes and choice-making action which has persisted through eons of rising suns.

But our understanding of what actually constitutes life is still not settled, and our current conclusions about what life entails may be deficient. For example, all cells are largely made of biochemical compounds known as proteins, as anthropologist Jeremy Narby described in *Intelligence in Nature*, (Narby, 2006, p. 144), and proteins are considered precursors to life as we know it on Earth (but not life itself). And yet, incredibly, proteins seem to exhibit a kind of intelligence![125] He quotes biochemist Christopher Miller, who, writing in the journal *Nature,* states: "Proteins are intelligent beings. They have evolved to operate in...a turbulent environment"). (Narby, 2006, p. 144).

Narby goes on to describe an interaction with Thomas Ward, a chemistry professor and a protein specialist, in which he asks whether Ward believes proteins have the capacity to know, and Ward replied, "A protein can move, powering itself from an external food source. A protein can interact with others of its own species, as well as with individual entities of other species, such as DNA and RNA molecules. A protein can build a large edifice, such as a cell. A protein can even reproduce itself, according to recent research. A protein can lose all of its functions, or 'die.' The foremost function of proteins is to recognize. For example, they recognize RNA molecules, or viruses, or other proteins. Then, based on this recognition, they can take appropriate

---

124 ...Perhaps even in entities we have previously not considered as living. Even the most basic of life forms, suggests Lipton, have a "primitive subjective awareness," as we saw. In fact, entities we have considered non-living may possess such a capacity, a notion that makes more sense when we recall Freeman Dyson's comment: "It appears that mind, as manifested by the capacity to make choices, is to some extent inherent in every electron," (Frankenberry, 2008, p. 372), or when we ponder the implication of this dawning reality that all that is is one interconnected being.

125 All beings that exhibit self-generated intelligence (the ability to acquire and apply knowledge and skills), are, by definition, considered to be alive. Each of Sahtouris' criteria for living systems is an aspect of this intelligence.

measures. If this is what you mean by 'to know,' then I find proteins undeniably have the capacity to know" (Narby, 2006, pp. 145-146).

A biochemical compound with the capacity to know?

Narby struggled to find an apt description for "intelligence," as the word has so many varied meanings and conceptualizations in our language. Finally, he settled on the expression "capacity to know" as best reflecting what we are trying to describe when we speak of "intelligence." What entities possess such a capacity? In the unity of all existence which we've been discovering, does the capacity to know exist throughout? Does the universe itself, the totality of existence, have the capacity to know?

The answer, quite rationally, appears to be yes. The universe, it seems, has *a vital essence*. Notice how many physicists, biologists, astrophysicists, cosmologists, molecular biologists, quantum theorists, engineers, anthropologists, neuroscientists, botanists, geneticists, mineralogists, philosophers, and others—a profusion of which we've noted in these pages—have pointed to this conclusion! In its beginnings were held the potentialities for life, consciousness, and self-reflective awareness, and for all that we see, live, think, feel, believe, relate to, experience, and all that exists within this present moment—even amid all the unpredictable and mind-boggling twists and turns of evolution.

The universe is a Vital Essence: a swirling, moving, ever-active, ever-evolving, ever-creating aliveness; an aliveness that is "able to reinvent itself in an evolutionary trajectory," and an aliveness that includes awareness and subjectivity, self-creation, complexity, communication, self-maintenance, mutuality, and creativity. It has the capacity to know. It is a universe that "must absorb information, store it, process it, and develop future strategies based on it." Our universe proceeds out of the capacity to know!

And Earth too consists of, fundamentally, the interaction between living and inert entities evolving greater and greater complexity in an interconnected system—a living unity with the capacity to know.

We are saying that the universe and everything in it, including Earth has evolved and deepened in complexity not by some undisclosed and mindless means, but rather out of an intelligence, a capacity to know, motivated by an energic and agentic force first of survival, and now, with evolving consciousness and self-reflective awareness, motivated by "thrival," an autopoietic striving toward an enhanced, deliberate quality of life. We are saying that what lives cannot be conceived of as having arisen from purely material substance, and that vitalism, "the theory that living organisms are truly alive, and not explicable in terms of physics and chemistry alone, not materialism," as Sheldrake states, "underlies our being."[126]

Sheldrake continues, "Mechanistic biology grew up in opposition to vitalism. It defined itself by denying that living organisms are organized by purposive, mind-like principles, but then reinvented them in the guise of genetic programs and selfish genes." (Sheldrake, 2012, p. 9). Then he observes the irony that "The dominant paradigm of modern biology, although nominally mechanistic, is remarkably similar to vitalism, with 'programs' or 'information' or 'instructions' or 'messages' playing the role formerly attributed to souls" (Sheldrake, 2012, p. 165).

And he reminds us that "before the seventeenth century, almost everyone took for granted that the universe is like an organism, and so was the earth. In classical, medieval and Renaissance Europe, nature was alive. Leonardo da Vinci (1540-1519), for example, made this idea explicit: 'we can say that the earth has a vegetative soul, and that its flesh is the land, its bones are the structure of the rocks… Its breathing and its pulse are the ebb and flow of the sea'" (Sheldrake, 2012, p. 28).

EARTHLIFE: ONE SINGLE UNFOLDING ENTITY

"Intelligence is the ability to adapt to change."

–Stephen Hawking (Redd, March 14, 2018, "Stephen Hawking").

Atoms and molecules, the stuff of matter such as it is, complexifies through interaction, sparked and spurred on by intelligent, informational, energic, and organic forces that reside in the invisible, dynamic space that

---

126 We're actually a composite ecosystem that's more bacterial than human, with one hundred trillion bacteria. The genetic material of humans actually consists of three genomes: one's parents; the epigenome, (change by turning genes on or off); and the DNA from your gut bacteria—called the microbiome ("a whopping 99 percent of your DNA"). Bacteria have always played an indispensable part in the phenomenon of life on Earth, observes Joan Borysenko (Howes, November-December 2016, "Food and").

fills the cosmos. At some early point, the unfolding process of the evolving universe complexifies into cells with the awakening ability, via the technology of DNA—complex organic molecules that saturate the universe—to self-create and self-maintain in the arena of the microscopic.

The pioneering scientist, James Lovelock, author of *Gaia: A New Look at Life on Earth,* (Lovelock, 2000), and consultant to NASA, noted that the living organisms present on the Earth have been responsible for regulating and generating the chemical balance of the air, of the seas, and of the land, so that life can continue to exist. The Gaia theory says that life forms on Earth create and maintain the specific environmental conditions necessary for these life forms to exist: autopoiesis in action, all connected.

Indeed, humanity is connected to all species. "*Other species are our kin,*" writes the influential biologist Edward O. Wilson "This statement is literally true in evolutionary time. All higher eukaryotic (nucleic) organisms, from flowering plants, to insects, to humanity itself, are thought of as having descended from a single ancestral population that lived 1.8 billion years ago. Single-celled eukaryotes and bacteria are linked by still more remote ancestors. All this distant kinship is stamped by a common genetic code and elementary features of cell structure. Humanity did not soft land into the teeming biosphere like an alien from another planet. We arose from other organisms already here" (Wilson & Kellert, 1993, p. 39).

With cellular self-creation and self-maintenance, a new level of information and creative possibility emerged out of the nascent capacities of genes, DNA, an apparent vessel for the manifesting of a capacity to know in our universe. "In the middle of the 1990's," says Narby, "biologists sequenced the first genomes of free-living organisms. So far, the smallest known bacterial genome contains 58,000 DNA letters. This is an enormous amount of information, comparable to the contents of a small telephone directory. When one considers that bacteria are the smallest units of life as we know it, it becomes more difficult to understand how the first bacterium could have taken form spontaneously in a lifeless chemical soup. How can a small telephone directory of information emerge from random processes?" (Narby, 1998, p. 42).

Narby, Sahtouris, Lipton, Wilson, Vertosick, Miller, Ward, and Lovelock, we should be clear, are talking about the unity of Earthlife within the context of an integrated yet ever-changing environment, manifested in the pervasiveness of complex organic molecules, DNA, on the planet. This environment, and all that it contains, is *one single interdependent entity* that is self-creative, holarchic, possessing self-knowledge, has the capacity to self-regulate or self-maintain, the ability to respond to stress, the capacity to transform matter, energy, and information, has communication among parts, empowered use of all parts, coordination among parts, is able to negotiate self-interest at all levels of hierarchy, exhibits reciprocity and mutual assistance of parts, has the ability to conserve what works well, and the capacity to change creatively. It is an autopoietic unity. Like the universe, our Earth has the capacity to know.

Sahtouris observes that "DNA is virtually the oldest thing in Earth's evolution still alive on its surface—propagating itself from the beginning in an unbroken chain, as surface rock transformed into endless creatures, who recycled it in turn into sediments that were subducted back into the magma of origin by great tectonic plates. All the while, DNA's species came and went, playing their roles and then disappearing, while it continued the dance" (Sahtouris, 2000, p. 381).

"Some biologists describe DNA as an 'ancient high biotechnology,' containing 'over a hundred trillion times as much information by volume as our most sophisticated information storage devices,'" writes biologist Robert Pollack, (Pollack, 1994, p. 28), ratifying Narby. This intelligence inherent in the DNA of all of Earth's life forms enabled the artistic brilliance of evolution on Earth to ingeniously burst, slowly at first, and yet eventually culminating in a creative living process which, in humanity, has transcended even its genetic intelligence itself through language and culture.[127]

---

127 "Recent work on genetic sequences is starting to reveal much greater complexity than could have been conceived even ten years previous to the data's emergence. How are scientists going to make sense of the overwhelming complexity of DNA texts? Robert Pollack proposes 'that DNA is not merely an information molecule, but is also a form of text, and that therefore it is best understood in analytical ways of thinking commonly applied to other forms of texts, for example, books.'" (Robert Pollack, 1997. "A Crisis of Scientific Morale," Nature. 385: 673-4). "This seems to be a sensible suggestion, but it begs the question: How can one analyze a text if one presupposes that no intelligence wrote it?" (Narby, 1998, p. 144).

Sahtouris notes that "Scientific research has shown for over half century, beginning with Barbara McClintock's work on corn plants, that DNA reorganizes intelligently in response to specific problems faced by living organisms. It happens in microbes and in very large multicelled creatures" (Sahtouris, 2000, pp. 61-62). The political science professor and prolific writer, Robert Wesson marvels: "No simple theory can cope with the enormous complexity revealed by modern genetics" (Wesson, 1991, p. 15).

"The black earth is alive with a riot of algae, fungi, nematodes, mites, springtails, enchytraeid worms, thousands of species of bacteria," says Wilson (Wilson, 1992, p. 335). This "riot" of life possesses immense quantities of information. "Biotechnology proves by its very existence the fundamental unity of life," writes Narby. "Each living being is constructed on the basis of the instructions written in the informational substance that is DNA. A single bacterium contains approximately ten million units of genetic information, whereas a microscopic fungus contains a billion units. In a mere handful of soil, there are approximately 10 million bacteria and one billion fungi. This means that there is more order, and information, in a handful of earth than there is on the surfaces of all the other known planets combined. The information contained in DNA makes a difference between life and inert matter" (Narby, 1998, p. 110).

Bruce Lipton, in the *Biology of Belief*, describes how this intricate, imaginative, intelligent, and *cooperative* vitality that is the understory of life itself has evolved: "For the first three billion years of life on this planet, the biosphere consisted of free-living single cells such as bacteria, algae, and protozoans. While we have traditionally considered such life forms as solitary individuals, we are now aware that signal molecules used by the individual cells to regulate their own physiologic functions, when released into the environment, also influence the behavior of other organisms. *Signals released by cells into the environment allow for a coordination of behavior among a dispersed population of unicellular organisms* [emphasis added]. Secreting signal molecules into the environment enhanced the survival of single cells by providing them with the opportunity to live as a primitive dispersed 'community'...

"Single celled organisms actually live in a community where they share their 'awareness' and coordinate their behaviors by releasing 'signal' molecules into the environment.

"To acquire more awareness and therefore increase their probability of surviving, cells started to assemble first in simple colonies and later into highly organized cellular communities" (Lipton, 2005, pp. 99-101). Even these primitive forms of life exhibited autopoiesis, the very feature which demonstrates the capacity to know.

"The construction of complexity begins with the biologically simple, one-celled organisms; the story is from the bottom up,' comments Rolston. "After certain thresholds are reached, parts can be differentiated. There is division of labor, a principal result of multicellularity. Now evolutionary development compounds the growth of biodiversity and biocomplexity" (Rolston, 2010, p. 54). It was a mere 700 million years ago—paltry moments in the life of the universe—that these communities of cells complexified into multicellular communities we call animals and plants and fungi.

"By tightly regulating the release and distribution of these function-controlling signal molecules," Lipton observed, "the community of cells would be able to coordinate their functions *and act as a single life form* [emphasis added] (Lipton, 2005, p. 101).

Narby underscores this point: "Even bacteria communicate. It turns out that all bacteria species relay information to one another in a 'bacterial Esperanto,' which they use to work together...Bacteria use chemicals rather than words to communicate, but this does not stop them from acting efficaciously" (Narby, 2006, p.143).

The predecessors of what becomes the brain first developed as the nucleated centers of single cells. Then, "as more complex animals evolved," Lipton continues, "specialized cells took over the job of monitoring and organizing the flow of behavior regulating signal molecules. These cells provided a distributed nerve network and central information processor...The brain controls the behavior of the body's cells.

"...Over a million years ago...multicellular organisms evolved into the first consciously aware *humans* to arrive on the scene. With consciousness, [self-reflective awareness], life was able to observe itself, reflect, and create its own future. Life could appreciate love and joy,"

Lipton effuses (Lipton & Bhaerman, 2009, p. xv). "Humans and a number of other higher mammals…evolved a specialized region of the brain associated with thinking, planning, and decision-making called the prefrontal cortex. This portion of the forebrain is apparently the seat of the 'self-conscious' mind processing" (Lipton, 2005, p. 104).[128]

We are describing a capacity to know—intelligence—that pervades all of life, with humanity representing the zenith of that intelligence on Earth. "Endowed with the ability to be self-reflective," says Lipton. "the self-conscious mind…can observe any programmed behavior we are engaged in, evaluate the behavior, and consciously decide to change the program. We can actively *choose* how to respond to most environmental signals" (Lipton, 2005, p. 104). And therein lies the motivator that transcends and includes survival alone: the motivation to thrive.

Lipton describes how communication results from consciousness, down to the level of primitive cellular life. The greater the sophistication, the richer the complexity of consciousness, the more powerful and the more impacting communication becomes; and so, the more the concept of "meaning" takes on richer forms. Yet, even among the earliest life forms on Earth, and continuing on through the evolutionary process that resulted in humanity, communication has a meaning. Earthlife has the capacity to know throughout. With the agentic capacity to know, intelligence, deepening over the history of evolution, richer strata of meaning result.

Evolutionary theorist John Stewart underscores the significance of the unity of Earthlife from the least to the greatest as being one holistic unified movement, citing the centrality of living evolutionary unfolding. Each "individual," large and small, is a part of one phenomenal progression:

"Evolutionary awareness shows us that it is an illusion to see ourselves or other individuals as distinct and separate entities. Individuals are inextricably part of an ongoing evolutionary process. They can have no existence without that process. They are born out of it, and can have

---

128 When Lipton talks about consciousness in this context, humanity's birth, he is not referring to the consciousness of other life forms. What he refers to as consciousness I am distinguishing as the rarer capacity exhibited in humans—and likely to some degree in a few other species—as self-reflective awareness, or native human consciousness. These are phenomena we will elaborate on in the pages ahead.

on-going effects only through what they return to it. In our mental processes, we can separate individuals out from the ongoing process and consider them as independent entities. But in reality they are never separate. Without an ongoing population that reproduces and evolves through time, there can be no individuals... Individual organisms such as ourselves are always parts of larger evolving processes." (Stewart, 2000, p. 316).

ANIMAL INTELLIGENCE

"The heavens may be clockwork, but the evolutionary epic is adventure."

—Holmes Rolston (Rolston, 2010, p. 55).

"Every living thing has a purpose, a mission, a life strategy, a set of gifts, and a set of weaknesses. Set aside any assumption that its behavior is random and meaningless."

—Jon Young, (Young, 2012, p. xxv).

"If a single cell of yellowy slime can solve a maze, does this not confirm that the entire edifice of life contains intelligence?"

—Jeremy Narby, (Narby, 2006, p.103).

Incredibly, right in front of us, is abundant and irrefutable highly documented and thoroughly researched evidence that species from every phylum of life, and arguably every species from every phylum of life exhibits a capacity to know (intelligence), an ability to make choices (agency), and an innate awareness (consciousness). Here, we survey the animal and plant kingdoms, citing example after example, in both primitive and complex species, of the ubiquity of intelligence, agency, and consciousness in all we think of as living.

A Pervasive Consciousness

"The common notion that humans alone experience consciousness is backward...Many animals are superhumanly alert... In 2012, scientists drafting the Cambridge Declaration on Consciousness concluded that 'all mammals and birds, and many other creatures, including octopuses have nervous systems capable of consciousness. (Octopuses use tools and solve problems as skillfully as apes—and they're *mollusks*,)," Carl Safina informs us in *Beyond Words: What Animals Think and Feel* (Safina, 2015, pp. 22-23).[129]

British psychologist and professor emeritus Max Velmans pointed out, "Only humans have full human consciousness...some non-human forms have unique nonhuman forms of consciousness...*even self-consciousness*" (Velmans, 2000, p. 281). Christof Koch, the head of the Allen Institute for Brain Science in Seattle writes, "Whatever consciousness is...dogs, birds, and legions of other species have it" (Safina, 2015, P. 23).

Jennifer Ackerman noted that, "for Darwin, even earthworms 'show some degree of intelligence' in their manner of dragging pine needles and vegetable matter to plug up their burrow" (Ackerman, 2016, p. 23).

Apparently, simple multicellular lifeforms each exhibit some capacity of consciousness.

Frankly, it is self-evident that every species of fauna as well as flora and fungi *knows* things, things we humans don't, and in every moment, utilizes their capacity to know in the service of survival. Every species knows what it must do to survive: on that level, large and small, every species must be conscious.

Further, sharing the same insight as that of Freeman Dyson, Graham Harvey asserts "Not only humans and elephants, but also rocks and subatomic particles act toward others. Their acting is not only performative of existing character and sociality, but also formative in the present unfolding of inherent potential for new possibilities" (Harvey,

---

129 Carl Safina is a writer, conservationist, and ecologist, whose writings have been recognized with numerous awards. His books include the award-winning *Song for the Blue Ocean, Eye of the Albatross, The View From Lazy Point; A Natural Year in an Unnatural World,* as well as *Beyond Words; What Animals Think and Feel,* which will be liberally referenced in this chapter

2006, pp. 202-203). Along with Dyson and Harvey, Sahtouris, Bohm, Elgin and a host of others agree, as we've seen: even rocks have some sort of existential awareness. They are an aspect of one interconnected consciousness.[130]

"Contrary to the damaging effects of Cartesian schizophrenia and the inherently hierarchical dualism of Newtonian physics and Lockean empiricism, there is plenty of scientifically acceptable evidence of consciousness in animals," writes Harvey. He continues, "ethologists recognize the cognitive competence of a wide range of other-than-human persons…Citing research about amoebas, anemones, snails and bats, Danah Zohar concludes, 'in varying degrees of quality and complexity, we can grant that all other animals are in some sense aware, capable to some degree of spontaneous and purposive activity, sensitive to stimuli something like pleasure or pain, and in possession of some rudimentary capacity to exercise free will. In the most primitive sense possible, possession of this set of qualities would also imply some sort of subjective "inner life" …every creature must have its own "point of view"' (Zohar, 1991, pp. 36-37). Harvey summarizes his view of personhood: "Analysts recognise personhood in other-than-humans and understand that to be a person is to be conscious and self-conscious, to act intentionally, with agency, and to communicate intelligently and deliberately" (Harvey, 2006, p. 187).

"Those scientists and philosophers who are willing to acknowledge that consciousness does not solely belong to humans may have escaped Descartes' anthropocentric dualism, but they have to face other difficult questions," Harvey suggests, wondering, "Where does consciousness come from? How far does it go: humans, animals, birds, insects, plants…What about rocks? Is matter conscious? If so, how did that happen? Has matter always been conscious or did consciousness emerge at some point in the process of evolution?" (Harvey, 2006, p. 189). As an animist, Harvey concludes that there is some form of awareness "all the way down" to rocks and even subatomic particles.

"If," marvels de Quincey, "the world is composed of matter that feels—of stuff that tingles with the spark of spirit all the way down

---

130 Given our materialist bias, it would see very easy to ridicule the idea that consciousness of some kind might go "all the way down." Yet with what we have already found, to be wise would be to remain open-minded!

to cells, molecules, atoms, and subatomic particles—then the world around us teams with sentient beings, themselves full of intrinsic meaning, constantly sending out messages as a kind of 'connective tissue' that unites all of nature in a universal network of shared, participatory meaning" (de Quincey, 2005, p. 91). We live in a world full of interconnected sentient beings, with messages as a sort of connective tissue, uniting all of nature in shared, intrinsic, and participatory meaning!

While appreciating that the notion of rocks and particles having some degree of existential awareness might be quite a reach for many, we do have to consider the intelligence and consciousness pervasive in nature, as we are about to document. All we experience as physical exists as one thing, and its complex and all-inclusive interplay precludes chunking out intelligent from not intelligent aspects.

Forethought, Learning, and Tool Use

One aspect of consciousness, and of intelligence, is forethought. Forethought involves decision-making and a capacity to know, in contrast to instinct, what Safina calls "an inborn pattern of activity," Ackerman's "learned innovative behavior."

Reflects Narby: "As I patrolled the texts of biology, I discovered that the natural world is teeming with examples of behaviors that seem to require *forethought*. Some crows manufacture tools with standardized hooks and toothed probes to help in their search for insects hidden in holes.

"Some chimpanzees, when infected with intestinal parasites, eat bitter, foul-tasting plants, which they otherwise avoid and which contain biologically active compounds that kill intestinal parasites.

"Some species of ants, with brains the size of a grain of sugar, raise herds of aphids which they milk for their sweet secretions and which they keep in barns. Other ants have been cultivating mushrooms as their exclusive food for fifty million years. It is difficult to understand how these insects do this without a formal consciousness" (Narby, 1998, p. 138).

"Insect tool use" marvels Safina, "is astonishing because it's so unexpected and looks so conscious. Various ants upon encountering liquid food such as rotting fruit, go away and return a minute later with leaves, sand grains, or soft wood to sop up the substance; then each ant can carry its weight in liquid food to the nest.

"Certain wasps use pebbles and earth to lock their prey into a hole along with the wasp's eggs (which will hatch and feed on the prey) ... "Assassin bugs hunt termites by first dressing in camouflage, sticking pieces of the termites' nest to their body so they smell like the nest...

"These are just a few examples of insect tool use," Safina observes. "Never mind the astounding construction, ventilation, food production, and heat-trapping functions of termite mounds, bee hives, spiderwebs, and the like. Does this mean that tool-using insects are very smart? Or does it mean that tool-making doesn't imply intelligence? Or is tool-making less impressive if tiny-brained insects do it? And how *about* those tiny brains: Are they aware? How aware? How are they making decisions, judging their progress? Does our own brain—as the science seems to show—come to a decision and then inform our conscious mind, making us merely believe we thought of it?" (Safina, 2015, pp. 197-198). [131]

Bees provide a good example of the complexity seen in lowly insects. "Now bees are no longer considered to be mindless automatons," reports Narby. They learn where food is and communicate about it. They *learn* (Narby, 2006, p.71). They manage abstract concepts (Narby, 2006, p. 55). They transfer knowledge across senses. "Though they have brains the size of pinheads, they can master abstract rules," says Narby ((Narby, 2006, p. 56).

Donald Griffin, biologist and "pioneer in the study of animal cognition" observes, "They are actually quite complicated...It seems to me, more likely than not, that there is some sort of continuum extending from the mental world of bees to us" (Narby, 2006, p. 45). Further, Griffin comments, "Certain highly social wasps can recognize individuals by their faces" (Safina, 2015, p. 22).

In the article entitled "Could a Bumblebee Learn to Play Fetch?" cognitive biologist Clint Perry described the research on animal

---

131 Robert Lanza notes, for example, as I have previously mentioned: "In 2008, experiments by Benjamin Libet and others...demonstrated that the brain, operating on its own, makes which-hand-to-raise choices that are detectible by observers watching brain-scan monitors up to ten seconds before the subject has 'decided' which arm to hold up." (Lanza, 2009, p. 197).

intelligence that he and his colleagues have been doing. "I want to know: How does the brain do stuff? How does it make decisions? How does it keep memory?" says Perry. And how big does a brain need to be in order to do all of those things? He introduced bumblebees to a puzzle requiring unfamiliar skills. "Perry and his colleagues wrote…in the journal *Science* that, despite bees' miniature brains, they can solve new problems quickly just by observing a demonstration. This suggests that bees, which are important crop pollinators, could in time adapt to new food sources if their environment changed," wrote the author Rae Ellen Bichell.

Bichell continues, "Perry and colleagues built a platform with a porous ball sitting at the center of it. If a bee went up to the ball, it would find that it could access a reward, sugar water…If a bee couldn't figure out how to get the reward, a researcher would demonstrate using a puppet… 'Bees that saw this demonstration learned very quickly how to solve the task. They got better over time,' says Perry.

"…The researchers found that bees that were able to watch a live bee do the trick first learned even faster… 'It wasn't monkey see, monkey do. They improved on the strategy that they saw,' says Perry. 'This all shows an unprecedented level of cognitive flexibility, especially for a miniature brain.'

Bichell expands: "…It's not just about the number of neurons in a brain, it's the connections between them. Research is showing more and more that animals—including tiny insects with 100,000 times fewer neurons than a human—can learn new skills quickly if their brains are wired right. Bees, for example, can count. They can make decisions by weighing uncertainty, and can even learn to pull a string in order to reveal a hidden cache of sugar water.

"'It's hard to see how it's an analog to anything they do in nature,' says Thomas Seeley, a biologist at Cornell University who has written books about bee behavior, including one on how honeybees make collective decisions" (Bichell, February 24, 2017, "Could a Bumblebee").

"The conventional notion that a bigger brain is always better and more powerful in vertebrates such as birds and mammals was finally put to rest," Ackerman notes, "with a simple but ingenious new way of measuring brainpower: counting neurons…The brains of birds may be small...but they 'pack surprisingly high numbers of neurons, *really* high, with densities at least akin to what we find in primates. And in corvids and parrots, the numbers are even higher'" (Ackerman, 2016, p. 54).

More specifically, Ackerman adds, "Much depends on where neurons are...elephant brains have three times the number of neurons found in the human brain (257 billion to our average 86 billion). But 98% of them are in the elephant cerebellum...An elephant's cerebral cortex, on the other hand, which is twice as big as ours, has only one-third the number of neurons found in our cerebral cortex... This suggests that what determines cognitive abilities is not the number of neurons in the whole brain but in the cerebral cortex" (Ackerman, 2016, p. 54), (and the connections between them, as Perry's research showed).

In all species, factors such as memories, decision-making, communication, and problem-solving are indicative of intelligence or the capacity to know, and thus are offering evidence of forethought. This type of intelligence differs from programmed, instinctive processes. Examples are abundant.

We have learned that cockroaches detect approaching predators by sensing minute air movements and possessing brain neurons that fire at a rate that varies with the wind, allowing them to deduce the direction of an attack. "A cockroach can perceive the world and take action in it, and its perception is inseparable from its sensorimotor capacities. It *knows* because it is *informed* by its body and brain about the approach of predator, and *embodies* action by scurrying away. This is no simple or reflexive process. The cockroach's nervous system decrypts the dynamics of minute air movements and sets in motion preventative action at the level of the whole organism. Just being a cockroach and coping with the world in order to stay alive requires *chi-sei* [the capacity to know]," declares Narby (Narby, 2006, pp. 141-142).

These types of behaviors lead to conclusions such as that of the noted British neurologist, naturalist, and author Oliver Sacks who remarked that, "It is increasingly evident that insects can remember, learn, think, and communicate in quite rich and unexpected ways" (Safina, 2015, p. 22).

Even invertebrates, like sponges, although lacking a brain and nervous system, "appear to make correct decisions on a regular basis," declares Narby. Octopi run mazes, open jars, disguise themselves, turn red and get angry, find food in concealed places, and transform their body shape, color, and skin texture. Further, "sea otters smash shellfish with stones while floating on their backs" (Narby, 2006, p. 71).

Moreover, incredibly, as we've seen, proteins move, fueling from external food sources, and can interact with others of its own species, as well as with other species, such as DNA and RNA molecules, can build a large edifice, such as a cell, can even reproduce themselves, and can lose all of its functions, or "die."

Even bacteria communicate, Lipton reminds us, using chemicals in lieu of words, which they use to work together in accomplishing their common goals.

Narby observes that a unicellular slime molds, part animal and part fungi, can consistently solve mazes, moving inward or outward by contracting and relaxing its gelatinous body in waves in a pattern that is self-organized. It has the ability to solve difficult mathematical problems. He comments: "A common view is that intelligence requires a brain. And brains are made of cells. But in the case of [the slime mold], a single cell [without nervous system or brain] behaves as if it had a brain." Further, he wonders, "If a single cell of yellowy slime can solve a maze, does this not confirm that the entire edifice of life contains intelligence?" (Narby, 2006, pp. 96-103).

Ratifying Ackerman's persuasive arguments on bird intelligence, Safina, too, states that birds "have a lot going on despite their much smaller brains. Especially wolf birds and their crow-family relatives, the jays, magpies, jackdaws, and rooks. They're *smart*. They're keenly observant and some share with dolphins, elephants, and certain carnivores a tool kit of reasoning, planning, flexibility, insight, and imagination—at an ape's level of intelligence," and notes that "New Caledonian crows can use tools to solve an eight-step puzzle to get at food.

"Cockatoos, which are parrots, also use insight in solving never-before-encountered puzzles involving locks, screws, and latches.

"Crows remember—for years—the faces of researchers who have caught and handled them for purposes of marking and measuring.

When they see those people walk across campus, they loudly scold them…Researchers have resorted to wearing masks or costumes when catching crows so that they won't be yelled at for years afterward…"

Further, Safina reports, "Two researchers wrote, 'New Caledonian crows and now rooks have been shown to rival, and in some cases outperform, chimpanzees in physical tasks, leading us to question our understanding of intelligence.' Scientists have concluded that, overall, ravens, crows, and their kin 'display similar intelligent behavior as the great apes.' New Caledonian crows and woodpecker (carpenter) finches use thorns to probe tree holes for bugs…" (Safina, 2015, pp. 192-196).

These representative examples of forethought, learning, and tool use, which seem to encompass the entire animal kingdom, persuasively argue that a capacity to know is inherent. Moreover, what has a capacity to know is conscious.

Cognition, Intelligence, and Consciousness

"Birds learn. They solve new problems and invent novel solutions to old ones. They make and use tools. They count. They copy behaviors from one another. They remember where they put things," summarizes Ackerman (Ackerman, 2016, p. 9).

"Of course, these are all human yardsticks of intelligence," she continues. "We can't help but measure other minds against our own. But birds also possess ways of knowing beyond our ken, which we can't easily dismiss as merely instinctual or hard-wired.

"What kind of intelligence allows a bird to anticipate the arrival of a distant storm? Or find its way to a place it has never been before, though it may be thousands of miles away? Or precisely imitate the complex songs of hundreds of other species? Or hide tens of thousands of seeds over hundreds of square miles and remember where it put them six months later? (I would flunk these sorts of intelligence tests as readily as birds might fail mine).

"Maybe *genius* is a better word" (Ackerman, 2016, p. 10).

"These days, however," Ackerman observes, if you "suggest that a bird has anything like human intelligence, consciousness, or subjective feeling, people might accuse you of anthropomorphizing, interpreting the behavior of a bird as if it were a human clothed in feathers. It's a natural human experience to project our own experience on the nature of other creatures, but this can—and does—lead us astray...

"Tempting as it maybe to interpret the behavior of other animals in terms of human mental processes, it's perhaps even more tempting to reject the possibility of kinship. It's what primatologist Frans de Waal calls 'anthropodenial,' blindness to the humanlike characteristics of other species. 'Those who are in anthropodenial,' says de Waal, 'try to build a brick wall to separate humans from the rest of the animal kingdom'" (Ackerman, 2016, pp. 22-23).

The bias against seeing the commonalities between humans and the rest of the animal kingdom, and the resistance to seeing the pervasive intelligence in, bluntly, every species, is, of course, based in the materialist posture that drives contemporary orthodox science and evolution theory. If everything is physical, material at its core, then consciousness, along with choice, subjectivity, and intelligence in other species must be denied or reconstituted, because these features are seen as epiphenomena, simply byproducts of matter. To conceive that these facets permeate nature is to challenge that premise, and instead have to reckon with the reality of these ubiquitous, primordial manifestations of consciousness.

"Consciousness is always *about* something.... In other words, consciousness always has a point," observes McIntosh. "In formal terms it is said to be *intentional*. Something is *intended* by it.... Intentionality is dynamic. Like a polarizing magnetic field that draws iron fillings into formations of multiple ellipses, consciousness aligns the processes of the mind into patterns with direction and purpose" (McIntosh, 2007, p. 260, taken from Combs, 2002, p. 8). Seeing consciousness in this way, it becomes self-evident that consciousness—intentionality—is pervasive throughout the natural world. Nature throughout has the capacity to know, which is what cognition is, and nature acts with intentionality based on cognition. Consciousness is demonstrated by cognition and by intentionality.

Because of the unconscious materialist bias so predominant in our science and our culture, says Ackerman, "most scientists who study birds [and other species as well] prefer the term *cognition* to *intelligence.* Animal cognition is generally defined as any mechanism by which an animal acquires, processes, stores, and uses information. It usually refers to the mechanisms involving learning, memory, perception, and decision-making. These are so-called higher and lower forms of cognition. For instance, insight, reasoning, and planning are considered high level abilities. Lower-level cognitive skills include attention and motivation" (Ackerman, 2016, pp. 23-24). Yet, if we use Narby's definition of intelligence, "capacity to know," then clearly what is being referred to as cognition is actually intelligence. A look at the dictionary shows that any of the same words are used interchangeably...and all involve obtaining and utilizing knowledge. The animal kingdom exhibits cognition *and* intelligence. Nature exhibits cognition and intelligence—a capacity to know— throughout, and so exhibits consciousness throughout.

Yet, not everything animals do is intelligent. Neither is everything humans do. Breathing is reflexive, for example. Digestion is not a conscious, intelligent process, but an autonomic one, as are most of our bodily processes. Narby comments on his conversation with Japanese biologist Toshiyuki Nakagaki. "In his view," Narby writes, "most internal information processing takes place on [an unconscious] level, even among human beings. 'I doubt anyone could explain how it is their body maintains balance while riding a bicycle. While we are riding, our body just naturally performs the calculations required to solve the equation'" (Narby, 2006, p. 105).

Ackerman gives a vivid example of how a reflexive process might look more intelligent than it is: "One striking example of this is cluster flocking—birds or other creatures moving in apparent unison, sometimes in large numbers...The great naturalist, Edmund Selous, who loved birds passionately and observed them with a scientific fervor, attributed this flocking phenomenon to telepathic thought transference from one bird to the next.... 'They must think collectively, all at the same time, or at least in streaks or patches—a square yard or so of an idea, a flash out of so of many brains'" (Ackerman, 2016, p. 29).

In southern Colorado, in the San Luis Valley, there is a protected area, the Monte Vista National Wildlife Refuge, where, every spring, tens of thousands of sandhill cranes assemble, and they perform a choreographed, charismatic, and spontaneous enactment of flight that *must—must—*be a symbiotic, telepathic phenomenon, a celebration of harmony and oneness. Different groups, for days, take off, fly irreproducible sky-bound bird-flight and then land gracefully on the many ponds: an annual, magnificent, one-off display of nature. I've seen it may springs now, and with greater, not less, marvel every time. Surely, Selous must have concluded correctly, intuitively, and wisely!

However, "We've since learned that the spectacular collective behavior of flocking birds (and schooling fish, herding mammals, swarming insects, and human crowds) is self-organized, emerging from simple rules of interaction among individuals. Birds are not 'transfusing thought,' communicating telepathically with their flock members to act in unison, as Selous surmised. Instead, each bird is interacting with up to seven close neighbors, making individual movement decisions based on maintaining velocity and distance with fellow flock members and copying how sharply a neighbor turns, so that a group of, say, four hundred birds can veer in another direction in a little over half a second. What emerges is almost instantaneous ripples of movement in what appears to be one living curtain of bird" (Ackerman, 2016, p. 30).

No less glorious and exhilarating to watch, but not group mind or telepathic communication. Something more ordinary, like, maybe, the Blue Angels, or a Nutcracker ballet, or a marching band on the football field. It is sometimes a challenge for researchers to determine whether a behavior they observe is an example of something highly intelligent or something only ordinary.

Still, intelligence can be seen everywhere: "Lately, the high, thin whistles and complex gargle calls of chickadees—the *fee-bees, zeees, dee-dee-dees,* and sibilant *stheeps*—have been parsed by scientists and declared one of the most sophisticated and exacting systems of communication of any land animal…chickadees use their calls like language, complete with syntax that can generate an open-ended number of unique call types…So reliable are chickadees vocalizations that other species heed their warnings" (Ackerman, 2016, p. 40).

"If cognition is defined as the mechanisms by which a bird acquires, processes, stores, and uses information," claims Ackerman, "then song learning is clearly a cognitive task: A young bird picks up information about how a song should sound by listening to a tutor of its own species. It stores this information in its memory and then uses it to shape its own song...scientists are noting the remarkable similarities of song learning in birds with human speech learning, from the process of imitating and practicing right down to the brain structures involved and the actions of specific genes...and the way song learning in a bird literally crystallizes brain structure, teaching us about the neurological nature of our own learning.

"'...The odd thing is, so many aspects of human speech acquisition are similar to the way that songbirds acquire their songs. In the great apes, there's no equivalent at all" (Ackerman, 2016, p. 139).

Ackerman concludes, "Some of our most complicated skills—language, speech, music—we learn the way birds do, through a similar process of imitation" (Ackerman, 2016, p. 141). "Songbirds learn their songs the way we learn languages, and pass these tunes along in rich cultural traditions that began tens of millions of years ago, when our primate ancestors were still scuttling about on all fours" (Ackerman, 2016, p. 4).

One more example of a startling capacity to know? The chickadee is "...also possessed of a prodigious memory. They stash seeds and other foods in thousands of different hiding places to eat later and can remember where they put a single food item for up to six months. "All of this with a brain roughly twice the size of a garden pea," comments Ackerman (Ackerman, 2016, p. 41), (And I still can't find my car keys!)

A recent study which illustrates the capacity to know found that, among horses in a herd, social signaling is important. Wendy Williams, a science journalist, equestrian, and author of *The Horse: The Epic History of Our Noble Companion,* commented that "Horses are highly social animals. In a natural state, they depend on each other for information that provides for the survival of the whole band. If a predator, for example, appears on the horizon, one horse immediately alerts the others through a wide variety of signals. Snorting, pricked ears and stamping are only a few of these signals. There's no reason why they wouldn't try to communicate with humans as well" (King, *NPR,* January 12, 2017).

Poignantly illustrating the degree of relationship and social connection in the lives of horses, a TV station in Phoenix, Arizona, recently showed a video and reported on a horse clan's reaction to the death of one of their mares: "Amazing recognition of death in wild horses. Sad, but beautiful…We did our very best today, to help a young wild mare whose baby had gotten stuck and died during delivery. Our experienced field team had jumped into action and our vet was getting there as fast as she could, but sadly the mare went into septic shock and passed, the baby had simply been stuck for too long. She was a beautiful dun mare, just 2 years old, her name was Clydette…Right after we moved away from her body, we witnessed how her band came and nuzzled her, after which the roan, her lead stallion, cried out for her very loudly. Shortly after that, they moved away from her body but stayed close. Other bands heard that call and suddenly came out of nowhere and then knew exactly where the lifeless body lied, even while there were no other bands around when she passed. What happened next was amazing; the other bands stood in line taking turns saying their goodbyes. First one band, then another. Then the two lead stallions of those two bands got into a short power struggle. Then you can see how Clydette's lead stallion comes running back one last time letting out a short scream in a last effort to protect her, or perhaps to tell everyone that she was his. It takes a most highly intelligent species to understand and actually mourn death. We have seen bands mourn their losses before, but for other bands to come and mourn her death also was simply awe inspiring. These animals have evolved to have amazing survival skills and very close and protective family bonds…." (Morell, September 21, 2016, "Horses can").

The results of the first research to investigate how horses respond to the state of knowledge or ignorance of their human companions, by Japanese scientists Monamie Ringhofer and Shinya Yamamoto of Kobe University, has been published online in the journal *Animal Cognition*. They note: "Previous studies have indicated that they are sensitive to bodily signals and the attentional state of humans…We found that horses communicated to their caretakers using visual and tactile signals. … These results suggest that *horses alter their communicative behaviour towards humans in accordance with humans' knowledge state*" [emphasis added] (Ringhofer & Yamamoto, December 15, 2016 "When horses").

Also, recently, we have learned that horses can use symbols to communicate with humans: this new study reported that "horses think carefully about what's going on around them. Horses are adept at learning and following signals people give them…The study's strong results show that the horses understood the consequences of their choices" [emphasis added] (Ringhofer & Yamamoto, December 15, 2016, "When horses").

Interaction, Theory of Mind, and Empathy

Safina spent time in Yellowstone National Park with Rick Macintyre who has spent *every day* of the last 15 years following its wolf population. Macintyre knows these wolves by temperament, personality, and predisposition. And he tells of their relationships and ways of being in the world:

"One of Twenty-one's favorite things was to wrestle with little pups. 'And what he really loves to do,' Rick adds, 'was pretend to lose. He just got a huge kick out of it.' Here was this great big male wolf. And he'd let some little wolf jump on him and bite his fur. 'He'd just fall on his back with his paws in the air,' Rick half-mimes. 'And the triumphant-looking little one would be standing over him with his tail wagging.

"'The ability to pretend,' Rick adds, 'shows that you understand that your actions are perceived by others. It indicates high intelligence.'

"…Twenty-one distinguished himself in two ways: he never lost a fight; and he never killed a vanquished wolf.

"Twenty-one's restraint in letting vanquished rivals go free seems incredible," marvels Safina. "What could it be? Mercy? Another term for a person who does not press their advantage against a threatening opponent is: magnanimous. Can a wolf be magnanimous?

"But do wolves have morals, ethics?

"Rick chuckles at the thought. 'It would be scientific heresy to say that they do, but—'" (Safina, 2015, pp. 144-147).

Wolves have a thriving relational and social life that involves interactive skills such as pretending, playing, empathy, perceiving the intent of others, anticipating how others think and what they might do, and showing mercy. Safina notes, "Strength impresses us, but we remember kindness... (Safina, 2015, p. 149).

"Rick offers. 'Imagine two groups of the same kind—two wolf packs, two human tribes, whatever. Which group is more likely to survive and reproduce: one whose members are more cooperative, more sharing, less violent with one another, or a group in which members are beating up and competing with one another?...' Minimal violence promotes group cohesion and cooperation. That's what a pack needs. The alphas set the example'" (Safina, 2015, p. 156). Adaptation, decision-making, cooperation, identity, innovation, intelligence. Consciousness.

"Wolves may not have words. What they have," affirms Safina, "is recognition, motivation, emotion, mental images, a mind map of their landscape, a roster of their community, a bank of memories and learned skills, and a catalog of scents with meanings attached to definitions. As we see in dogs, that's easily more than enough to understand who's who and what's where in a lifetime" (Safina, 2015, p. 211). Still, wolves and dogs are only two of untold species of animals that exhibit consciousness.

"Consciousness seems necessary," states Safina, "when we must judge things, plan, and make decisions" (Safina, 2015, p. 21). "To understand elephants, we must delve into topics like consciousness, awareness, intelligence, and emotion..." (Safina, 2015, p. 20). These qualities seem to be surprisingly magnified when we discover the richness in elephants' consciousness, and in their relationships with each other.

"Dhruba Das, who trains Indian villagers to reduce human-elephant conflict, has commented, 'It's more like wisdom. They can sense things. They know what to do. They'll take whatever a situation offers them and use it to their best advantage'" (Safina, 2015, p. 36).

"Elephants form deep social bonds developed through deep time. Parental care, satisfaction, friendship, compassion, and grief didn't just suddenly appear with the emergence of modern humans. All began their journey as pre-human beings." Safina observes, "Our brain's provenance is inseparable from other species' brains in the long cauldron of living time. And thus, so is our mind" (Safina, 2015, p. 30).

"Elephants seem better than apes—even humans—at keeping immediate track of a large number of individuals. Their recognition ability exceeds primates...When researchers played a recorded call of an absent family member or bond group member, elephants returned the call and moved toward the sound. Playing a recording of an elephant outside their bond group, they didn't react noticeably. But when played calls of total strangers, they bunched defensively, raising their trunks to smell...'Intelligent, social, emotional, personable, interactive, respectful of ancestors, playful, self-aware, compassionate...' wrote Cynthia Moss, [who has studied elephants in Africa for forty years], along with Joyce Pool [elephant communication expert], and several colleagues" (Safina, 2015, p. 38).

"To elephants and some others, it matters who has died. It's why they are 'who' animals. The importance of memory, learning, and leadership is why individuals matter. And so. A death matters to the survivors," says Safina.

"...A researcher once played a recording of an elephant who had died" reported Safina. "The sound emanated from the speaker hidden in a thicket. The family went wild calling, looking all around. The dead elephant's daughter called for days afterward. The researcher never again did such a thing" (Safina, 2015, p. 67).

It is difficult to watch elephants' relationships with each other and not to see the deep connection they feel, and the degree of compassion they can show to their own kind. "Researchers once saw an elephant pluck up some food and place it in the mouth of another whose trunk was badly injured. 'Elephants show empathy,' Amboseli researchers Richard Byrne and Lucy Bates state plainly. They aid the ailing. They *help* one another" (Safina, 2015, p. 61).

Humans are able to experience a sense of self and see that there are other humans who also have a sense of self. And we distinguish our self from that of others. Humans have a theory of mind.

Likewise, wolves, dogs, horses, elephants—as we've seen—along with dolphins, apes, some birds, and a whole host of other animals have a theory of mind. They demonstrate that they can understand others' thoughts and motives, and that other beings think differently—and they relate to and interact on that basis. "Theory of mind...is an idea," Safina states. "Naomi Angoff Chedd, who works with autistic children, tells me it is 'knowing another can have thoughts different than yours.' I like that definition. Dolphin researcher Diana Reiss says it the ability to feel that 'I have an idea what's on your mind.' That's different. Still others assert—oddly I think—that it's the ability to read the mind of others...." (Safina, 2015, p. 244).

Safina says, with some frustration, "So far, some scientists grant theory-of-mind ability—basically, understanding that another can have thoughts and motives that differ from yours—to apes and dolphins. A few allow elephants and crows. Occasional researchers have admitted dogs. But many continue to insist that theory of mind is 'uniquely human.' Even while I was writing this, science journalist Katherine Harmon wrote, 'On most animal species, scientists have failed to see even a glimmer of evidence.'"

Continues Safina, incredulous: "Not a glimmer? It's *blinding*. People who don't see the evidence aren't paying attention" (Safina, 2015, pp. 245-246).

So animals of the same species in many cases demonstrate "theory of mind," a sort of relational eavesdropping.

"Even more surprising...than interspecies eavesdropping," says Jon Young, "is inter-order and even interclass eavesdropping, which many...studies confirm (Young, 2012, p. 102).

"Frans de Waal [primatologist and ethologist] pays attention." Safina argues, citing de Waal's observation of the interspecies communication and playfulness of chimps. "The shenanigans of chimps who like to spray water on unsuspecting zoo visitors, he says, reflects 'a complex, and familiar, inner life... Watching free-living animals negotiate the world on their terms shows you their rich mental abilities'" (Safina, 2015, p. 24).-246).[132]

---

132 When Safina talks about free-living animals he is distinguishing between "wild" animals (and plants) and those that are "domesticated," an important distinction. "Domesticated' means genetically changed from wild ancestors by selective breeding...zoos have wild animals in captivity; farms have domesticated animals. Arboretums have wild plants; farms have domesticated plants. 'Domesticated' doesn't mean tame. A wolf that has been captive-born and bottle-raised and is entirely tame is a wolf in captivity: it's not *domesticated*. Pet parrots, even captive-bred, are not domesticated.

"Domestication implies intentional human creation of animal and plant varieties and breeds that don't exist in nature. Classically that's accomplished by selective breeding, but now technicians use genetic engineering, too. Farmers, fanciers, and researchers select traits they want and specifically promote them. They breed individuals with those traits together, resulting in many varieties

A Universe Full of Magical Things

When humans interact with dogs, we can see that dogs process both meaning and intonation. "Dogs process both what we say and how we say it in a way which is amazingly similar to how human brains do," says Attila Andics, a neuroscientist at Eotvos Lorand University in Hungary—meaning in the left hemisphere of the brain, and intonation in the right hemisphere. "Humans seem to be the only species which uses words and intonation for communicating emotions, feelings, inner states," he says. "To find that dogs have a very similar neural mechanism to tell apart meaningful words from meaningless sound sequences is, I think, really amazing."[133]

Great apes and their fossils have been studied to try to determine when that shift to left brain dominance occurred. "It seems the story is really that there is a general mammalian bias to process words or meaning in communication in the left hemisphere and it became exaggerated in humans," says Brian Hare, a cognitive neuroscientist at the Duke Institute for Brain Sciences in Durham, N.C. "It's not something completely new to our species."

"Scientists believed that our unique language abilities evolved out of human's left hemisphere dominance in processing communication," states Hare, "This really challenges that, because dogs also have a left hemispheric bias for processing words with meaning. That's a surprise."[134]

Coyotes too exhibit left hemisphere dominance—they too are members of the dog family, North American canids, Canidae. "In Native American mythology the coyote is often depicted as a trickster,

---

of domestic chickens, cows, pigs, pigeons, and laboratory rats, terriers, farmed salmon, corn, rice, wheat and so on—all genetically changed from their naturally evolved wild ancestors. Dogs' social skills are their wolf heritage, but *dogs' orientation toward humans* resulted from domestication...Dogs' friendliness results from their genetically altered brain chemistry. Ours does too," (Safina, 2015, pp. 221-225).

133 According to Jon Young, in *What the Robin Knows,* "The abstract of a piece about...research on monkeys and baboons published in *The Annals of the New York Academy of the Sciences* in 2006 states: "Historically, a dichotomy has been drawn between the semanti communication of human language and the apparently emotional calls of animals. Current research paints a more complicated picture. Just as scientists have identified elements of human speech that reflect the speaker's emotions, field experiments have shown that the calls of many animals provide listeners with information about objects and events in the environment. Like human speech, therefore, animal vocalizations simultaneously provide others with information that is both semantic and emotional" (Young, 2012, pp. 104-105).

134 Their Masters' Voices: Dogs Understand Tone and Meaning of Words on *NPR* August 30, 2016, Nell Greenfield Boyce.

using its wit and cunning to outsmart its foes. So how smart are coyotes, really?" asks the website www.KnowYourNeighbors.net. "While it's difficult to compare animal intelligence to ours, taking a look at the coyote's incredible range of abilities points us towards some sort of answer. Coyotes are the most vocal North American canids... with over 11 unique vocalizations. These include growls, yips, and those distinctive howls you may have heard cutting through a still Nevadan night. While this verbal repertoire helps them communicate when living in packs, coyotes also have the ability to live alone or in pairs.

"A coyote's intelligence and ability to learn is on full display when hunting and scavenging." Says the website, "Rattlesnake carcasses have been found at coyote dens picked clean, save for the head; other complete rattlesnake carcasses have been found with bite marks behind the neck, suggesting that coyotes may either teach their pups how to properly kill the deadly serpents, or at least how to eat them. Coyote pairs have been observed hunting porcupines together, learning through trial and painful error how to flip them over and avoid those dangerous quills. Coyotes have even been spotted hunting prairie dogs alongside badgers! Researchers observed that coyotes benefited more from these team-ups than did badgers, reinforcing that deviously scheming stereotype of coyotes" (http://www.knowyourneighbors.net).[135]

---

135 The treatment of coyotes, bobcats, wolves, black bears, red foxes, cougars, and other American predators might lead us to conclude either that these animals have no capacity to know—or that their capacity to know is exponentially greater than ours. The January 25th, 2016 issue of *High Country News* in an article by Ben Goldfarb entitled "The Forever War," documented that "In 2014, Wildlife Services exterminated 796 bobcats, 322 wolves, 580 black bears, 305 cougars, and 1,186 red foxes. And that's nothing compared to coyotes. That year, the agency killed 61,702, one coyote every eight and a half minutes...Even 2014's eye-popping coyote kill total represented the agency's lowest figure in more than 20 years."

We might need to reexamine who it really is that has a capacity to know altogether: according to an article in *The Guardian* "Humanity has wiped out 60% of animal populations since 1970, report finds," by Damian Carrington, Environment editor on October 29, 2018, "Humanity has wiped out 60% of mammals, birds, fish and reptiles since 1970, leading the world's foremost experts to warn that the annihilation of wildlife is now an emergency that threatens civilisation. The new estimate of the massacre of wildlife is made in a major report produced by WWF [World Wildlife Federation] and involving 59 scientists from across the globe. It finds that the vast and growing consumption of food and resources by the global population is destroying the web of life, billions of years in the making, upon which human society ultimately depends for clean air, water and everything else.

"If there was a 60% decline in the human population, that would be equivalent to emptying North America, South America, Africa, Europe, China and Oceania. That is the scale of what we have done. "This is far more than just being about losing the wonders of nature, desperately sad though that is," he said. "This is actually now jeopardising the future of people. Nature is not a 'nice

Safina writes of his two dogs, Chula and Jude: "The dogs chase each other in totally unnecessary play. They fake each other out...Chula will try to double back to intercept Jude, but Jude will stop to see from which way Chula is coming. They know what is going on, and they seem to understand that the other is trying to fool them. That's 'theory of mind'... One is evaluating what the other is thinking, each showing a clear understanding that the other might be faked into a false belief about which direction they'll be charging from. Because they're playing, there's both cleverness and humor in this. (Unless they're just two unconscious machines interacting without sensation or perception. Some people insist that 'we can't be sure.' That's what I mean by denial.)

"A dog who has never before seen a ball would not bring it to a person and lay it at their feet. But a dog experienced with balls comes to invite play. They envision a game, plan a way to start it, and execute the plan with a human partner who they understand is knowing. Theory of mind (Safina, 2015, p. 248).

"While we tend to lack a theory of *their* minds, other animals seem to have a theory about ours. They know that we know," contends Safina (Safina, 2015, p. 256).

"At Gombe Stream National Park in Tanzania, a researcher watched two adult male chimpanzees separately climb on top of a ridge at sunset. There they noticed and greeted each other, clasped hands, sat down together, and watched the sun descend. Another researcher wrote of a free-living chimpanzee gazing for fifteen minutes at an especially striking sunset. If they are really admiring the sunset, it's probably no deeper reason than that it looks pretty to them. Same as us" (Safina, 2015, p. 55).

"Chimpanzees understand both the goals and intentions of others as well as the perception and knowledge of others," reports Safina. But what they tend to do with that information can be very different than what an elephant might do: "With no training, chimpanzees can cooperate to pull ropes together to retrieve a heavy box of food. But they seldom do...Chimpanzees lack dogs' human-like skills because

to have' – it is our life-support system...

"Other recent analyses have revealed that humankind has destroyed 83% of all mammals and half of plants since the dawn of civilisation and that, even if the destruction were to end now, it would take 5-7 million years for the natural world to recover..."

chimpanzees lack dogs' human-like cooperative tendencies" (Safina, 2015, p. 264).

"Chimps are jealous, ambitious, and frequently aggressive with their own group. Chimpanzee groups are male dominated. Male chimps form coalitions against other males...A bonobo group's dominant individual is never a male, always a female. Female coalitions dominate, preserve the peace, and keep males socially submissive. Female authority dampens male aggression."

Safina continues, musing, "Ever wonder why, when chimps are often so nasty, bonobos are so friendly and sexy with one another? ...Bonobos evolved after the Congo River formed, about a million years ago, isolating a population of chimpanzees south of the river. Somehow, for the bonobos, a lot changed" (Safina, 2015, p. 260).

"Compared to chimpanzees, bonobo brains have more gray matter in regions involving perceiving distress in others. Bonobos have a larger nerve pathway for controlling aggressive impulses, inhibiting harm toward others. This limits stress, dissipates tension, and reduces anxiety to levels that open up room for sex and play... compared to chimpanzees, bonobos mature more slowly physically, psychologically, and socially, and they learn skills more slowly. The *same* genes that lowers aggressiveness by creating a more juvenile-like brain chemistry also create more juvenile physical features."

"As chimps reached adulthood, they become less playful and largely intolerant of sharing. Bonobos are like chimps who never quite grow up...." (Safina, 2015, p. 229-231).

"Darwin coined the term 'natural selection' because he was comparing the mechanics of what happens in nature with the artificial selection applied in raising livestock," Safina explains. "But nature doesn't really select: it filters. The environment works as a filter, and as the environment changes, it filters differently. The point is: as the pressures change, we remain a work in progress" (Safina, 2015, p. 236).

Interspecies communication by definition assumes forethought, a capacity to know, and the ability to influence and be influenced in communication. We can see that the higher up the developmental ladder, the greater the capacities, on a variety of levels, a given species may have. But we need to be careful in our assumptions of what "skills" any given species may be capable of demonstrating.

For example, Carl Safina asks the question, "You might already have taken in stride that dogs and apes indicate and understand intent. But what about, say, a fish?

"We think of apes as smart, because apes *are* smart—*and* because they look like us. But the science of cognition is becoming peppered with reports of 'apelike performance' in some other animals. Newest among them: certain fish. Our rarified list of beings who use gestures to direct the attention of companions—humans, bonobos, dolphins, ravens, African hunting dogs, wolves, domestic dogs—must now include groupers. Yes, the same fish that's in countless fried fillet sandwiches; they're among the smartest…

"Flexible interspecies cooperation of the sort shown by groupers and their partners is so rare, even humans engage in it with only two or three species" (Safina, 2015, pp. 258-260).

Intelligence in Farm Animals

What about intelligence, the capacity to know, in farm animals? "From pigs to cows, sheep to chickens, farm animals are all much smarter than we've ever given them credit for. Pigs learn their names and can do tricks like a dog. Cows, goats, and chickens all have incredibly complex social constructs, and they have best friends just like we do. These are all amazing, sentient beings," says Anna Vallery on the "One Green Planet" website (Vallery, "Farm animals"). The website highlights some of these species' exceptional skills:

Pigs: "Researchers at the University of Cambridge found that not only do pigs recognize themselves, but they also show an understanding of how mirrors work, and can use the reflections to find food… As if that wasn't enough to convince you that pigs are incredibly intelligent, they are also known to play games (in exchange for a delicious treat). Pigs like to play with toys, such a balls, and are prone to getting bored if they aren't provided with enough stimulation."

Cows: "Cows have extremely good memories. It has been found that they not only recognize faces, but they will remember faces even after a long period of time. Cows also remember where to find the best grazing spots and directions to their favorite watering hole… Cows even have a social hierarchy among the members of their herd. There is typically one cow who is the "boss" and dictates the behavior of her followers. If a cow doesn't want to listen to this head cow, they are isolated from the herd (just like high school). And when a new cow is introduced to the herd, she has to network and build relationships with other members of the herd before she is fully accepted."

Chickens: "Chickens are amazing mothers and take care of their babies long before they have hatched. It has been found that they 'talk' and 'purr' to the eggs during incubation. When the chicks hatch, hens are even more loving. They defend their babies from predators, show empathy for their chicks, and teach their young everything they need to know… When chicks are hatched, moms continue to teach them all the ways of the chicken. They teach them what is safe to eat and what to avoid. They also teach them about the social hierarchy, or pecking order…Baby chicks are pretty brilliant right from the beginning. They are known to show object permanence, the ability to understand an object exists, even when they can't see it. Chicks develop this ability when they are around two days old, while it take human babies six months to learn this skill."

Sheep. "Sheep are capable of recognizing all kinds of faces. They recognize sheep in their flock and are aware when these sheep are missing. They can recognize "bully" sheep, and get distressed when they come around. These sheep can even recognize the person who cares for them and the sheepdog that herds them! If the appearance of another individual is altered, the sheep have no problem still identifying who it is, and they can keep track of over 50 different sheep faces!"

Goats: "Researchers from Queen Mary University of London and the Institute of Agricultural Science in Switzerland always suspected that there was more to goats than meets the eye and found that goats are excellent at puzzles…These researchers presented goats with a puzzle, originally intended for primates, and placed food inside a box that can only be reached by solving the puzzle. The goats had to use their teeth to pull on a rope to activate a lever, and then lift the lever up with their muzzle."

Safina noted that "psychology professor Gordon Gallup…has said,

'Self-awareness provides the ability to contemplate the past, to project into the future, and to speculate on what others are thinking'" (Safina, 2015, p. 273). By this definition—a reasonable one—every animal we've discussed fits the bill: all are self-aware; thus, all have the capacity to know. Safina observes that self-recognition is why a wolf eating an elk's leg doesn't bite into its own leg. A concept of "self" is absolutely basic (Safina, 2015, p. 278).

Individuality and Personality

The relational capacities in myriad species are also seen in orangutans and dolphins: "Orangutans can evaluate how well a human is understanding their gestures. When gestures fail, they sometimes pantomime what they would like from a human... Orangutans are able to establish shared meaning if the humans prove capable of understanding what they're trying to express" (Safina, 2015, p. 292-293).

"Self-recognition, capacity to know, ability to communicate, even across species, and to maintain these relationships over time, to recognize individuals, to play—these are all skills that are highly developed in dolphins. They call each other by name, and they answer when they hear their own name called," notes Safina (Safina, 2015, p. 313). "It seems extraordinary that these free-living creatures view humans as worthwhile playmates. That they do carries big implications about minds understanding minds. Here they are, and this is *who* they are, coming across that species bridge on their own terms, bringing their own invitation, offering their own game, playing by their own rules. They've done this many times (Safina, 2015, p. 334).

Dolphins' communication emphasizes relationship and connection to a degree that is greater than even we humans. "For the most part," notes de Quincey, "humans speak to share information and details about ourselves and our environment—more often than not to enhance our ability to do something, to manipulate some part of our world. It is *instrumental* communication. By contrast, we learn that the cetaceans [dolphins, porpoises, whales] communicate primarily to restore, enhance, or create an experience of harmony and well-being both within the pod and between the pod and the ocean around them" (de Quincey, 2005, p. 258).

"Partway into considering examples of dolphin 'cognition,' I realized that dolphins are as cognizant as anyone," Safina says. "There are so many examples of them acting aware and clever (because they are aware and clever) that one might as well try to compile examples of humans acting aware and clever. That's just who we are. And it's who dolphins are, too. Dolphins and humans have not shared a common ancestor for tens of millions of years. Yet for all the seeming estrangement of lives lived in liquid, when they see us they often come out to play, and we greet them and can recognize in those eyes that someone very special is home. 'There is someone in there. It's not a human, but it is someone,' says Diana Reiss" (Safina, 2015, p. 336).

Safina notices that certain animals actually have their own individual personality. "Personality is probably the most under-recognized aspect of free-living creatures." He says. "Dolphins have personality galore. They're born with personality. Shy. Bold. Rambunctious. Bullying" (Safina, 2015, p. 392).

"Paul Spong, a psychologist who'd worked at the Vancouver Aquarium, has written, 'Eventually my respect verged on awe. I concluded that *Orcinus orca*, [the largest member of the dolphin family, and so-called "killer whale,"] is an incredibly powerful and capable creature, exquisitely self-controlled and aware of the world around it, a being possessed with a zest for life and a healthy sense of humour and, moreover, a remarkable fondness for and interest in humans.'"

Safina challenges: "If that seems a little, well, anthropomorphic— that's the point" (Safina, 2015, p. 398).[136]

---

136 For those interested in the debate in science as to whether we are imposing our anthropomorphic (human-oriented) perspective onto animals, and, in ways, therefore, seeing them as humanlike with human skills and motivations and awarenesses, take a look at Chapter 2 and 3 in Safina's *Beyond Words*. Traditional science has denied consciousness to all but humans, characterizing all other species as instinctive, that is, "natural" or *unlearned*, having "an inborn pattern of activity or tendency to action, a natural or innate impulse, inclination, or tendency."

This stands in contrast to a perspective that recognizes the developmental continuum we share with other species, which suggests an *intuitive approach*, "easy to understand or operate without explicit instruction" or an *insightful approach*, "apprehending the true nature of a thing, especially through intuitive understanding, penetrating mental vision or discernment; a faculty of seeing into inner character or underlying truth, an understanding of relationships that sheds light on or helps solve a problem, an understanding of the motivational forces behind one's actions, thoughts, or behavior; self-knowledge." (Dictionary.com). This intuitive or insightful view toward living beings suggests innovation, creativity, and deliberate selection within an environmental context, an intentional evolution.

"*When individuals matter*—when you're a 'who'—you need a social brain capable of reasoning, planning, rewarding, punishing, seducing, protecting, bonding, understanding, sympathizing," notes Safina. "Your brain needs to be your Swiss army knife, packing different strategies for different situations. Dolphins, apes, elephants, wolves, and humans face similar needs: know your territory and its resources, know your friends, monitor your enemies, achieve fertilization, raise babies, defend, and cooperate when it serves you" (Safina, 2015, p. 346).

"We see stereotypes when we see 'elephants' or 'wolves' or 'killer whales' or 'chimps' or 'ravens.' But the instant we focus on individuals, we see that individuals differ. We see an elephant named Echo with exceptional leadership qualities; we see wolf named Seven fifty-five struggling to survive the death of his mate and exile from his own family; we see a lost whale who is lonely but humorous and stunningly gentle. It's not *person*ality; it's *individual*ity. And it's a fact of life. And it runs deep. *Very* deep," remarks Safina (Safina, 2015, p. 392).

"Researchers have also published findings on the individual personalities of rats, mice, lemurs, finches and other songbirds, bluegill and pumpkinseed sunfish, stickleback fish, killifish, bighorn sheep, domestic goats, blue crabs, rainbow trout, jumping spiders, house crickets, social insects….In other words, pretty much everywhere they've looked, they've found that individuals differ. Some are more aggressive, bolder, shier, some more active, some fear the new, while others are explorers… Judy [Weiss, of Rutgers University] elaborated: 'We really don't appreciate how much personality animals have. Even as scientists, we've hardly ever thought about that'" (Safina, 2015, p. 395).

Biologist Julie Young, of the Predator Research Facility near Logan Utah, noted about carnivores, "Carnivores have personalities. Some are going to cause problems…Their behavioral profiles fall along this bell curve on the bold-shy spectrum. Too shy and you're not going

---

Similarly, by the way, traditional science has denied intelligence to non-human species, opting instead for an attribute of "cognition," as we have seen. Intelligence is something attributable to humans; cognition to animals, in many cases, I would say, as a way to avoid anthropomorphizing. But in *EarthDance*, Sahtouris takes the perspective that we are better off "anthropomorphizing" than "mechanomorphizing;" "For scientists who shudder at such *anthropomorphism*—defined as reading human attributes into nature—let us not forget that *mechanomorphism—reading mechanical attributes into nature*—is really no better than second-hand anthropomorphism, since mechanisms are human products. Is it not more likely that nature in essence resembles one of its own creatures than that it resembles in essence the non-living product of one of its creatures?" (Sahtouris, 2000, p. 10).

to establish territory. Too bold, and you're probably getting shot" (Goldfarb, January 25th, 2016, "The Forever").

Ackerman makes the point strongly: "Birds do have personalities. Some scientists shy away from the term with its anthropogenic overtones, preferring temperament, coping style, behavioral syndrome. But call it what you will, individual birds behave in ways that are stable and consistent across time and in different circumstances, just as we do. There are the bold and the meek, the curious and the cautious, the calm and the nervous, the fast learners and slow learners...

"The study by [researcher Lucy] Alpin's team not only revealed affiliations between birds with similar personalities. It also found that the bolder birds flit between groups expanding the size of their social networks and enhancing their access to food sources" (Ackerman, 2016, pp. 110-111).

"The animal kingdom is symphonic with mental activity, and of its millions of wave-lengths, we're born to apprehend the minutest sliver," concludes the reverent Safina: (Safina, 2015, p. 307). "There is no more wondrous fact than that we are kin, bee and bird of paradise—and great elephant—stardust, all" (Safina, 2015, p. 57).

All of nature is living, conscious. All of earth is nature.

PLANT INTELLIGENCE

And what about the capacity to know in nature's flora?

"'Can a plant be *intelligent?*' asks Michael Pollan, in the *Science Friday* web article, "New research on plant intelligence may forever change how you think about plants. Some plant scientists insist they are—since they can sense, learn, remember and even react in ways that would be familiar to humans...

"For the longest time, even mentioning the idea that plants could be intelligent was a quick way to being labeled 'a whacko.' But no more, which might be comforting to people who have long talked to their plants or played music for them" (Pollan, 2013, "The intelligent").[137]

---

137 Michael Pollan, author of *The Botany of Desire, The Omnivore's Dilemma,* and others, discusses with *Science Friday,* (hosted and produced by Ira Flatow), his 2013 article in the *New York Times,* entitled "The Intelligent Plant: Scientists debate a new way of understanding flora." Pollan was named in

Pollan reiterates what we have seen, that "philosophers and psychologists have been arguing over the definition of intelligence for at least a century, and whatever consensus there may once have been has been rapidly slipping away. Most definitions of intelligence fall into one of two categories. The first is worded so that intelligence requires a brain; the definition refers to intrinsic mental qualities such as reason, judgment, and abstract thought. The second category, less brain-bound and metaphysical, stresses behavior, defining intelligence as the ability to respond in optimal ways to the challenges presented by one's environment and circumstances. Not surprisingly, the plant neurobiologists jump into this second camp.

"'I define it very simply,' [Italian plant physiologist Stefano] Mancuso said. 'Intelligence is the ability to solve problems.' In place of a brain, 'what I am looking for is a distributed sort of intelligence, as we see in the swarming of birds'…

"Mancuso's hypothesis is that something similar is at work in plants, with their thousands of root tips playing the role of the individual birds—gathering and assessing data from the environment and responding in local but coordinated ways that benefit the entire organism.

'Neurons perhaps are overrated,' Mancuso said. "They're really just excitable cells.' Plants have their own excitable cells, many of them in a region just behind the root tip."

"The new research, he says, is in a field called plant neurobiology—which is something of a misnomer, because even scientists in the field don't argue that plants have neurons or brains.

"[Plants] have analogous structures," Pollan explains. "They have ways of taking all the sensory data they gather in their everyday lives … integrate it and then behave in an appropriate way in response. And they do this without brains, which, in a way, is what's incredible about it, because we automatically assume you need a brain to process information" (Pollan, 2013, "The intelligent").

Incredibly, plants, too, have intentions, make decisions, and compute complexities within their environment: "Science now indicates

the 2010 *TIME* 100 annual list of the world's 100 most influential people, and has numerous other awards as well. He has written copiously for 25 years on the relationship between nature and culture.

that plants, like animals and humans, can learn about the world around them and use cellular mechanisms similar to those we rely on. Plants learn, remember, and decide, without brains," says Jeremy Narby (Narby, 2006, p. 91). He observes, "Individual plant cells also appear to have a capacity to know" (Narby, 2006, p. 141).[138]

Pollan reasons, "The hypothesis that intelligent behavior in plants may be an emergent property of cells exchanging signals in a network might sound far-fetched, yet the way that intelligence emerges from a network of neurons may not be very different. Most neuroscientists would agree that, while brains considered as a whole function as centralized command centers for most animals, within the brain there doesn't appear to be any command post; rather, one finds a leaderless network. That sense we get when we think about what might govern a plant—that there is no there there, no wizard behind the curtain pulling the levers—may apply equally well to our brains."

"Plants have evolved between fifteen and twenty distinct senses, including analogues of our five: smell and taste (they sense and respond to chemicals in the air or on their bodies); sight (they react differently to various wavelengths of light as well as to shadow); touch (a vine or a

---

138 Christian de Quincey elaborates on Jeremy Narby's career experiences and his resulting conclusions, pointing toward a perspective on plants' and on human's potential relationships with them. Narby's work goes a long way toward helping us in the western world make sense of how shamans have been able to discover the various healing powers of the plant world that we now use indispensably in western medicine today: "Jeremy Narby studied anthropology at Stanford University and among the Quirishari people in the Amazon rain forest of Peru," says de Quincey. "His central message is that shamanic knowledge, gained from direct communication with certain hallucinogenic plants, reduced by centuries if not millennia modern scientific data about the basis of life. He knows that shamanic visions of entwined or double serpents, recorded throughout history and across the world, foreshadowed in remarkable detail the twentieth-century discovery of the double-helix or twisted letter structure of DNA the fundamental molecule of life.

"Many shamans use special plant guides—such as *ayahuasca, peyote,* and psilocybin mushrooms, along with special chants, whistles, drumming, or dancing—to prepare the mind for receiving altered states of consciousness capable of penetrating alternative realities.

"How had Narby gathered his understanding of Western molecular biology—particularly knowledge of DNA—and the visionary insights of the shamans? After some shrewd detective work, he discovered that the key had to be photons. He came to believe that the visionary dreams shamans told him about were actually conversations with spirits contained in hallucinogenic plants—and these conveyed accurate knowledge about DNA and the role of light in all living systems...

"He tracked down evidence from different sources in science, mythology, anthropology, and shamanism, paying attention to the most unlikely clues. Then, using a technique called the *focalizing* learned from a shaman guide, he started putting the pieces together. You began to see *forms* in the teaching of the various disciplines. Most dramatic of all the clues was a form common to both the scientific description of DNA is actually a twisted ladder, or double helix, and the shamanic description of the essence of life as entwined cosmic serpents" (de Quincey, 2005, p. 233-234).

root "'knows' when it encounters a solid object); and, it has been discovered, sound," all of which assist in foraging for food, competing for resources with other plants, and fending off predators, observes Pollan (Flatow, January 9, 2014, "New research"). Some plants signal for help, releasing chemicals that discourage predators, and communicate cellularly using electrical and molecular signals. They can detect nearby greenery, alter the direction and even shape of their stems, and regulate their growth and development in adjusting to changing environment (Narby, 2006, pp. 83-94).

"With no apparent nervous system, *plants make the same chemicals*—such as serotonin, dopamine, and glutamate—that serve as neurotransmitters and help create mood in animals, including humans," notes Safina. "And plants have signaling systems that work basically as do animals', though slower...Plants sense and respond to chemicals in the air, soil, and on themselves. Plants' leaves turn to track the sun. Growing roots approaching an obstacle or toxin, sometimes alter course *prior* to contact. Plants have reportedly responded to the *recorded* sound of a munching caterpillar by producing defensive chemicals...

"But except for insectivores and sensitive-leafed plants, most plants behave too slowly for the human eye" (Safina, 2015, p. 23).

"Scientists have...found that the tips of plant roots, in addition to sensing gravity, moisture, light, pressure, and hardness, can also sense volume, nitrogen, phosphorus, salt, various toxins, microbes, and chemical signals from neighboring plants. Roots about to encounter an impenetrable obstacle or a toxic substance change course before they make contact with it. Roots can tell whether nearby roots are self or other and, if other, kin or stranger," reports Pollan (Flatow, January 9, 2014, "New research").

Ground ivy senses nutrient rich verses nutrient poor locations, and grows toward the richer patches. "An individual plant has enormous capacity for changing its morphology, its branching structure, to accommodate the environment in which it finds itself," says Narby. "The transformation occurs very slowly from a human point of view, over a period of months, rather than milliseconds" (Narby, 2006, pp. 83-87).

The parasitic plant called dodder moves by wrapping itself around other plants, and correctly assessing their nutritional quality, and then grows more or less coils, thus exploiting more nutrient rich plants. "It has to make a correct decision or face death," foraging strategies that rival those of animal foragers, without the advantage of a brain, exhibiting foresight, plasticity, intelligence, and adaptation (Narby, 2006, p. 88).

Pollan summarizes the perspective of those who conclude that agency and the capacity to know are features of flora as well: "…proponents believe that we must stop regarding plants as passive objects—the mute, immobile furniture of our world—and begin to treat them as protagonists in their own dramas, highly skilled in the ways of contending in nature. They would challenge contemporary biology's reductive focus on cells and genes and return our attention to the organism and its behavior in the environment. It is only human arrogance, and the fact that the lives of plants unfold in what amounts to a much slower dimension of time, that keep us from appreciating their intelligence and consequent success. Plants dominate every terrestrial environment, composing ninety-nine per cent of the biomass on earth. By comparison, humans and all the other animals are, in the words of one plant neurobiologist, 'just traces'" (Pollan, 2013, "The intelligent").

Pollan continues, opining, "Once the definition of 'behavior' expands to include such things as a shift in the trajectory of a root, a reallocation of resources, or the emission of a powerful chemical, plants begin to look like much more active agents, responding to environmental cues in ways more subtle or adaptive than the word 'instinct' would suggest. 'Plants perceive competitors and grow away from them,' Rick Karban, a plant ecologist at U.C. Davis, explained, when I asked him for an example of plant decision-making. 'They are more leery of actual vegetation than they are of inanimate objects, and they respond to potential competitors before actually being shaded by them.' These are sophisticated behaviors, but, like most plant behaviors, to an animal they're either invisible or really, really slow.

"The sessile life style also helps account for plants' extraordinary gift for biochemistry, which far exceeds that of animals and, arguably, of human chemists," asserts Pollan. "(Many drugs, from aspirin to opiates, derive from compounds designed by plants). Unable to run away, plants deploy a complex molecular vocabulary to signal distress, deter or poison enemies, and recruit animals to perform various services for them" (Pollan, 2013, "The intelligent").

"Perhaps the most troublesome and troubling word of all in thinking about plants is 'consciousness,'" writes Pollan. "If consciousness is defined as inward awareness of oneself experiencing reality—'the feeling of what happens,' in the words of the neuroscientist Antonio Damasio—then we can (probably) safely conclude that plants don't possess it. But if we define the term simply as the state of being awake and aware of one's environment—'online,' as the neuroscientists say—then plants may qualify as conscious beings, at least according to Mancuso and Baluška" (Pollan, 2013, "The intelligent").

What about larger flora, such as trees? In ways, trees are the most fascinating flora of all. "Trees in a forest organize themselves into far-flung networks, using the underground web of mycorrhizal fungi which connects their roots to exchange information and even goods," observes Pollan. "This 'wood-wide web,' as the title of one paper put it, allows scores of trees in a forest to convey warnings of insect attacks, and also to deliver carbon, nitrogen, and water to trees in need" (Wohlleben, 2015, p. 47).

They are capable of learning, so they must store experiences they can draw on somewhere. Says Peter Wohlleben, in *The Hidden Life of Trees*, "If trees are capable of learning (and you can see they are just by looking at them), then the question becomes: Where do they store what they have learned and how do they access this information? After all, they don't have brains to function as databases and manage processes."

The answer, Wohlleben believes, is a tree's roots. The roots of the tree are comparable to the brain of a human. They are, he says, "the most important part of the tree. Conceivably, this is where the tree equivalent of the brain is located. Brain, you ask? Isn't that a bit far-fetched? Possibly, but now we know that trees can learn. This means they must store experiences somewhere, and therefore, there must be some kind of a storage mechanism inside the organism. Just where it is nobody knows, but the roots are the part of the tree best suited to the task...

"It is now an accepted fact that the root network is in charge of all chemical activity in the tree... Roots absorb substances and bring them into the tree. In the other direction, they deliver the products of photosynthesis to the tree's fungal partners and even route warning signals to neighboring trees. But a brain? For there to be something we would recognize as a brain, neurological processes must be involved, and for these, in addition to chemical messages, you need electrical impulses. And these are precisely what we can measure in the tree…"

Wohlleben notes that "In conjunction with his colleagues, František Baluška, from the Institute of Cellular and Molecular Botany at the University of Bonn is of the opinion that brain-like structures can be found at root tips. In addition to signaling pathways, there are also numerous systems and molecules similar to those found in animals. When a root feels its way forward in the ground, it is aware of stimuli. The researchers measured electrical signals that led to changes in behavior after they were processed in the 'transition zone.' If the root encounters toxic substances, impenetrable stones, or saturated oil, it analyzes the situation and transmits the necessary adjustments to the growing tip. The root tip changes direction as a result of this communication and steers the growing root around the critical areas" (Wohlleben, 2015, pp.82-83).

Wohlleben's perspective, and the perspective that seems to be the inarguable conclusion of our conversation here, is that flora, like fauna, learn, alter their behaviors based on circumstance and situation, and make decisions. They possess consciousness. Wohlleben concludes: "The distinction between plant and animal is, after all, arbitrary, and, [in addition to capacity for autonomous movement or not], depends on the way an organism feeds itself: the former photosynthesizes and the latter eats other living beings Finally, the only other big difference is the amount of time it takes to process information and translates it into action" (Wohlleben, 2015, p. 83).

"'Plants can do incredible things. They do seem to remember stresses and events…They do have the ability to respond to 15 to 20 environmental variables,' Pollan says. 'The issue is, is it right to call it *learning*? Is that the right word? Is it right to call it *intelligence*? Is it right, even, to call what they are *conscious*? Some of these plant neurobiologists believe that plants are conscious — not self-conscious, but conscious in the sense they know where they are in space…and react appropriately to their position in space…

"'So intelligence may well be a property of life,' speculates Pollan. 'And our difference from these other creatures may be a matter of difference of degree rather than kind. We may just have more of this problem-solving ability and we may do it in different ways" (Flatow, January 9, 2014, "New research").

Chapter Nine

———∾∾———

EVOLUTION—EMERGENCE,
AGENCY, AND INNOVATION

David Stenhouse, a New Zealand philosopher and psychologist, described intelligence as "adaptively variable behavior within the lifetime of the individual" (Narby, 2006, p. 85). Using this understanding of what intelligence is, or Narby's "capacity to know," the intelligent conclusion is that intelligence is ubiquitous in nature.

"Science now indicates that plants, like animals and humans, can learn about the world around them and use cellular mechanisms similar to those we rely on," asserts Narby." Plants learn, remember, and decide without brains.

"There is really nothing equivalent to living beings made of living cells," he says. "Each individual cell in a body is alive. Living cells are themselves creatures with a life cycle, and they must look after their own survival by adapting to the circumstances they encounter," (Narby, 20056 p. 125), (pointing, it should again be noted, to the interconnectedness of all of nature. As de Quincey reminds us, "Nothing can happen anywhere in this total energy matrix without it affecting the whole. The universe, it seems, is a cosmic quantum") (de Quincey, 2002, p. 71).

Narby offers this reflection: "How had we come from Darwin, who admired the mental faculties of ants, to this? I turned this question over in my mind for days. It was as if most Western biologists had fallen into a mechanical trance for most of a century, out of which they were only just emerging. I did not fully understand why things had happened this way. But I did feel relieved that science was changing and revealing intelligence in nature once again. And this confirmed some of the most ancient beliefs of indigenous people…

"Science has gone through profound changes in recent decades, and scientists are starting to argue against sacrosanct principles such as Occam's razor," Narby believes.[139] "Some scientists are realizing that there is little evidence that nature is simple, or that simple accounts are more likely than complex ones to be true" (Narby, 2006, pp. 52-53).

If we take a look at our current materialist perspective on evolution, we will see that with the presumption of the primacy of matter, we cannot begin to comprehend or appreciate the intelligence and the wisdom that saturates the natural world. Such a simplistic understanding cannot lead us to a meaningful view of our world and our universe.

"Contrary to what conventional science and religion have been telling us, evolution is neither predictable nor pre-determined, but rather, an intelligent dance between organism and environment," states Lipton. "When conditions are ripe—either through crisis or opportunity—something unpredictable happens to bring the biosphere into a new balance at a higher level of coherence" (Lipton & Bhaerman, 2009, p. xvi).

Evolution is neither predictable nor pre-determined, but rather, an intelligent dance between organism and environment!

"Neo-Darwinism insists that random accident and natural selection are the sole 'mechanisms' of evolution," observes Sahtouris. "Yet the self-organized creatures and ecosystems—habitats—such as that which… [evolve] through the genetic information exchange web of bacteria…are not readily explained as simple accumulations of lucky accidents. Nor does natural selection amount to a real theory, [as we'll

---

139 Occam's razor or Ockham's razor is a philosophical principle attributed to William of Ockham (1287–1347), which argues that the simplest hypothesis, with the fewest assumptions, should be considered to be most likely. Science uses the idea to further develop theoretical models, but it is not a notion consistent with the scientific method. It values simplicity as a means of arriving at truth, and is often preferable to more complex possibilities because they are more readily testable.

see, Karl Popper discusses below], since it tells us little more than that some creatures die before they reach the age of reproduction" (Sahtouris, 2000, pp. 109-110).

Rather than being a mindless mechanism, we will see that evolution is a self-creating and self-maintaining process, an autopoietic process that has the capacity to know, and is self-aware, living, becoming, and unfolding, as one fabric. It is a process that flows from a universe that contains, of course, the very features that are emerging: aliveness, creativity, intelligence, consciousness, and self-reflectiveness. These features don't emerge or evolve, or awaken (!) out of nothing—to claim that they do asks a grander miracle than science ought ever to feel comfortable with. No, these features are fundamental, inherent, and integral to the universe itself, and they define it and give it meaning. The process is vital: as Elgin says, it is the Life Force. Sahtouris, Elgin, Narby, Lipton, de Quincey, and a broad swath of other researchers, scientists, and others from numerous and wide-ranging disciplines and experience—as we've seen—have described aspects of life and how it likely emerged. It is an emergence out of the universe's mighty potential, not at all merely a modifier of matter.

As we have discussed above, the traditional scientific view, the view that has dominated western cultures for 350 years, is that the universe, (according to the scientist Newton and the philosopher Descartes), is a grand machine, a collection of parts composed of matter, functioning in a cause and effect way. As we've seen, there is no dynamic life process here, only mechanics only "natural" selection and mutation errors. There is no awareness or emotion or growth or self-creation; no aliveness. Atoms have no inherent life in this conventional model, and yet, in contradiction, when matter evolves in complexity (inspired by what causality is not quite clear), life and mind in general and self-reflective human beings in particular result.

All being in a lifeless universe is conventionally understood in terms of matter and cause-and-effect mechanics, and yet, inexplicably, life, mind, consciousness, and self-reflective consciousness result—emerging, so the story goes, as epiphenomena, mere byproducts of chemical reactions with inert and dead substances. Consciousness is considered to be, with no evidence, (since science has no empirical data about it), simply a manifestation of this somehow emerging biology, located in the brain.

But this view, which is actually only an article of mechanistic science's religious faith—dogma not data—is constructed out of the unproven and undocumented assumption that all is based in physical, and so all is matter or a secondary effect of matter. Now, however, many others are beginning to voice a different view, as we are seeing.

For example, Christian de Quincey references David Bohm, saying, "Bohm's cosmology is radical when viewed from the perspective of mainstream science: He believed that a complete theory of the cosmos must take in consciousness. This must be so because clearly consciousness is an undeniable reality of the universe. Without it we would know absolutely nothing" (de Quincey, 2002, p. 163). The same can be said for what we call life, as it too is an undeniable reality in the universe. If intelligence, consciousness, and life *exist* in the universe, these qualities must be *inherent* in the universe. They are not, and cannot be merely epiphenomena: they are intrinsic, essential, vital, integral, innate, elemental, and natural.

It is nothing short of remarkable that an outdated science with so many gaps and so little explanation, evidence, and documentation would be so earnestly clung to by so many for so long—despite the vast amounts of new knowledge *its own disciplines* have gathered over the last one hundred and twenty years.

Now, this outdated scientific dogma is being everywhere tested and challenged and with it a new order of knowledge is unfolding. *The universe is growing deeper in its capacity to know.* We have arrived at notions about the world that are far different than the science of the last three hundred and fifty years. In fact, "science has entirely overturned what we know about the structure of the world," says the physicist Evan Harris Walker. "But rather than revising our picture of what reality is, we cling to a collage of incongruent shards. We preserve a false assemblage of images, one pasted upon another, so that we can keep unchanged the mental portrait of ourselves and the worlds to which we are accustomed" (Walker, 2000, p. 7).

The great fear is that, should we expand our view, and truly wrestle with being, life, mind, intelligence, consciousness and self-reflective awareness and intentionality, and love, compassion, ideas, heart, or soul, that we'll be cast with no compass into a world of magic, superstition, and make-believe: non-science. So, instead, we ordain that everything that is—the quantum, the zero-point field, the energy of space, of the universe, of the big bang, persons and personality, love, intimacy, relationship, Spirit, joy, pain, the love of baseball—all are artifacts of lifeless, inert matter. Clanging and clanking and dead? Or Spirit infused in nature and in the universe? Which is more likely to be make-believe?

EVOLUTION: REDUCTIONIST SCIENCE OR SENTIENT PROCESS?

Bruce Lipton, in *Spontaneous Evolution,* describes the materialistic paradigm and its assumptions as it relates to evolutionary thought in this way: "Scientific materialism has offered four tenets in the dominant basal paradigm that, until recently, have been accepted and regarded as indisputable scientific fact:

1)Only Matter Matters—the physical world we see is all there is.

2)Survival of the Fittest—Nature favors the strongest individuals, and the Law of the Jungle is the only real natural law.

3)It's in Your Genes—we are victims of our biological inheritance and the best we can hope for is that science finds a way to compensate for our inherent flaws and frailties.

4)Evolution is Random—life is basically random and purposeless, and we got here pretty much the same way as an infinite number of monkeys pecking on an infinite number of typewriters over an infinite amount of time might produce the works of Shakespeare" (Lipton & Bhaerman, 2009, p. 87).[140]

"The old story—the paradigm of mechanism and materialism— was Charles Darwin's theory of evolution first published in 1859,"

---

140 This is a reference to what is called the "infinite monkey theorem," which, according to Wikipedia, states that "a monkey hitting keys at random on a typewriter keyboard for an infinite amount of time will almost surely type a given text, such as the complete works of William Shakespeare," A " probability of a universe full of monkeys typing a complete work such as Shakespeare's *Hamlet* is... *extremely* low (but technically not zero)...In this context, 'almost surely' is a mathematical term with a precise meaning, and the "monkey" is not an actual monkey, but a metaphor for an abstract device that produces an endless random sequence of letters and symbols."

explains de Quincey. "Darwin said that all the varieties of animals and plant species populating the Earth were the result of chance variations subjected to the process of natural selection. Combining Newtonian mechanism-materialism with Darwinian evolution, neo-Darwinians argued that all life forms were the sole result of chance mutations in the blind selection of a dumb and blind natural 'watchmaker.'

"New species emerged in evolution through random genetic mutations—spontaneous reshuffling of molecules in the genes—and these novelties survived if they happened to improve the host organism's ability to produce offspring. 'Survival of the fittest' meant the survival of those offspring with genes for producing further offspring well adapted to a particular ecological niche. It was all a matter of chance and necessity—the randomness of mutations channeled by the determinism of mechanism in the interplay between organism and environment. There was no room for purpose in nature. All progress, all evolution, was due merely to the injection of novelty through mutations, selected according to the necessities of survival in a particular niche...

"Consciousness, when it finally arrived late in the game, was merely an epiphenomenon, an ineffective byproduct of matter that had evolved to the complexity of nervous sytems and brains" (de Quincey, 2002, pp.20-21).

A plethora of highly respected scientists with various academic credentials, as we are seeing, have challenged these reductionist assumptions with an alternative paradigm of a vital, intelligent, deliberate, communicative, cooperative, agentic, holistic, and organic universe and evolutionary *process*—a universe that experiences awareness throughout.

Despite the difficulties the rubric of science has in reckoning with this dynamic aliveness—that its own facts repeatedly document—the problem with understanding our existence as if it were a giant, mindless, lifeless machine, and the ludicrousness of that stance, has become unequivocally clear. Writes Jeremy Narby:[141]

---

141 In this section, I cite Jeremy Narby's arguments in *The Cosmic Serpent*, (Narby, 1998), and reference a number of the authors and experts he has referred to in that book. I do so to echo the very important point he is making—that an evolutionary theory that is mechanistic is highly inadequate, false, and detrimental to a post-reductionistic, quantum-inspired, and life-inspired scientific perspective. I am grateful for his thoughtful and valuable writing.

Modern biology…is founded on the notion that nature is not animated by intelligence and therefore cannot communicate.

This presumption comes from the materialist tradition established by the naturalists of the eighteenth and nineteenth centuries. In those days, it took courage to question the explanations about life afforded by a literal reading of the Bible. By adopting a scientific method based on direct observation and the classification of species, Linnaeus, Lamarck, Darwin, and Wallace audaciously concluded that the different species have evolved over time.…

Wallace and Darwin simultaneously proposed a material mechanism to explain the evolution of species. According to their theory of natural selection, organisms present slight variations from one generation to the next, which were either retained or eliminated in the struggle for survival. This idea rested on a circular argument: those who survive are the most able to survive. But it seemed to explain both the variation of species and the astonishing perfection of the natural world, as it has retained only the improvements. Above all it took God out of the picture and enabled biologists to study nature without having to worry about a divine plan within…

It wasn't until 1950s and the discovery of the role of DNA that the theory of natural selection became generally accepted among scientists. The DNA molecule seemed to demonstrate the materiality of heredity to provide the missing mechanism. As DNA is self-duplicating and transmits its information to proteins, biologist concluded that information cannot flow back from proteins to DNA; therefore genetic variation can only come from errors in the duplication process. Crick, termed this the "central dogma" of the young discipline called molecular biology. "Chance is the only true source of novelty," he wrote (Crick, 1981, p. 58).

The discovery of DNA's role in the formation in molecular terms of the theory of natural selection gave a new impetus to materialist philosophy. It became possible to contend on a scientific basis that life was purely material phenomena. Francis Crick wrote: "the ultimate aim of the modern movement in biology is to explain *all* biology in terms of physics and chemistry" (Crick, 1966, p. 10).

...The materialist approach in molecular biology... rested on the unprovable presupposition that chance is the only source of novelty in nature, and that nature is devoid of any goal, intention, or consciousness (Narby, 1998, pp. 132-134).

The cultural giants who first examined the theory of evolution presumed that intelligence was not a feature to be considered. Instead they proposed a mechanistic notion of natural selection, a process wherein the organisms (both flora and fauna) which are better adapted to their environment are the organisms that more likely will survive and reproduce. Organisms vary to small degrees from generation to generation, with occasional mutations, which, in time, result in evolved features and even new species. In general, notes Narby, the most desirable features are the ones that were retained over generations, and these are the features most often selected.

Sheldrake observed that "The neo-Darwinian theory of evolution...differed from Charles Darwin's theory in that *it rejected the inheritance of acquired characteristics.* Instead, *organisms inherit genes* from their parents, passing them on unaltered to their offspring, unless there were mutations, that is to say, random changes in the genes... In Neo-Darwinism, the thread of life is literal: helical DNA molecules in thread-like chromosomes dispense to mortals their destiny at birth" [emphasis added] (Sheldrake, 2012, pp. 42-43).

The DNA molecule was discovered in the 1950s and thought to be the mechanism of natural selection, which produced genetic variation by self-duplicating and transmitting DNA's information to proteins—and it was further reasoned that genetic variation can *only* come from errors in this duplication process, the so-called "central dogma" of evolution theory. Natural selection was purely the result of chance! This perspective reinforced the materialistic conclusion that "life was purely material phenomena," and that "nature is devoid of any goal, intention, or consciousness."

And yet, chance doesn't seem at all to be the operative principle among at least some contemporary researchers. Recall Narby's conversations with biochemist Christopher Miller, (who wrote in the journal *Nature,* "Proteins are intelligent beings. They have evolved to operate in...a turbulent cellular environment"). and chemistry professor and protein specialist Thomas Ward "(A protein can move, powering itself from an external food source. A protein can interact with others of its own species, as well as with individual entities of other species, such as DNA and RNA molecules. A protein can build a large edifice, such as a cell. A protein can even reproduce itself, according to recent research. A protein can lose all of its functions, or 'die.' The foremost function of proteins is to recognize. For example, they recognize RNA molecules, or viruses, or other proteins. Then, based on this recognition, they can take appropriate measures. If this is what you mean by 'to know,' then I find proteins undeniably have the capacity to know"). Further, these behaviors don't seem to have much to do with chance.

Thus, contrary to this reductionist dogma, Sahtouris concludes that "It is certainly obvious by now that DNA can reorganize itself and repair the kinds of accidental change that was thought to be the only way to evolution. It is a relief to know we are not just piled up accidents and copying mistakes, but beings who have organized and evolved ourselves in harmony with other living beings that form our environment. It is good to know that life is too intelligent to proceed by accident" (Sahtouris, 2000, p. 116).

Also contradicting this *undocumented and unproven* materialistic view of evolution that unfortunately continues to dominate conversations on the subject, biologist Ernst Mayr, who is both an authority on the subject of evolution, and considered to be a leading evolutionary biologist, sometimes credited with inventing the modern philosophy of biology, in 1982 pointed out that there is *"no clear evidence for any change of the species into a different genus or for the gradual origins of an evolutionary novelty"* [emphasis added] (Mayr, 1982, pp. 529-530).

And Sheldrake points out that "There is a vast gulf between rhetoric about the powers of genes and what they actually do...Thanks to the discoveries of molecular biology, we know what genes actually do. They code for the sequences of amino acids that are strung together in polypeptide chains, which then fold up into protein molecules. Also, some genes are involved in the control of protein synthesis.

He asserts, "DNA molecules are molecules. They are not 'determinants' of particular structures, even though biologists often speak of genes 'for' structures or activities, such as genes 'for' curly hair or 'for' nest-building behaviors in sparrows... They merely code for the sequences of amino acids in protein molecules" (Sheldrake, 2012, p. 163).

Further, Narby directly challenges "the unprovable presupposition that chance is the only source of novelty in nature, and that nature is devoid of any goal, intention, or consciousness," citing a 2005 *Time* magazine article by J. Madeline Nash in which she raises "another fundamental problem [that] contradicts the theory of chance driven natural selection. According to the theory, species should evolve slowly and gradually, since evolution is caused by the accumulation and selection of random errors in the genetic text. However, the fossil record reveals a different scenario" (Narby, 1998, p.141).

She writes in a paleontology review of the research: "Until about 600 million years ago there were no organisms more complex than bacteria, multicelled algae, and single cell plankton.... Then, 543 million years ago, in the early Cambrian, within the span of no more than 10 million years, creatures with teeth and tentacles and claws and jaws materialize with the suddenness of apparitions. In a burst of creativity like nothing before or since, nature appears to have sketched out the blueprints for virtually the whole the animal kingdom.... Since 1987, discoveries of major fossil beds in Greenland, and China, in Siberia, and now in Namibia have shown that the period of biological innovation occurred at virtually the same instant in geological time all around the world....,... Virtually everyone agrees that the Cambrian started almost exactly 543 million years ago, and even more startling that all but one of the phyla in the fossil record appeared within the first 5 to

10 million years" (Nash, December 5, 1995, "When Life").

The piecemeal pace of change that mechanistic evolutionary theory espouses seems unpersuasive, in light of the abrupt, transformative, and global alteration of nature's landscape in the Cambrian period. If eight out of nine phyla of life arose within an ecological moment of a mere ten million years and across the planet—in contrast to the pace of transformation in other epochs and locales—the change in the pace of the selection and mutation process would have to have been accounted for other than by customary explanations that attribute such speed to a rich, abundant ecological environment. Rapid evolutionary change doesn't seem to correlate in other lavish environments, except perhaps in small, isolated environments like the Galapagos Islands. Perhaps the rapid transformation of the entire evolutionary landscape of Earth in such a brief period of time has more to do with the intelligence and agency that we have seen demonstrated by a range of species making informed choices based in their capacity to know!

"Throughout the fossil record," states Narby, "species seem to appear suddenly, fully formed and equipped with all sorts of specialized organs, then remain stable for millions of years" (Narby, 1998, p. 141).

In 1972, Stephen Jay Gould and Niles Eldredge proposed the concept of "punctuated equilibrium" to describe in part the process of evolutionary change, an idea that lends support to the Darwinian model of evolution. Darwin had concluded that the pace of evolution was slow with small variations over an extended period of time. However, if evolution is gradual, we should be able to find evidence for incremental changes in the fossil record, and yet, for the most part, we haven't.

Changes occur at the edges of species' populated areas, according to this notion of punctuated equilibrium, where it is more likely that small groups will be separated from the primary population, and undergo changes to survive in their new environment, in time establishing a new, non-interbreeding species, such as with the Galapagos Islands (PBS, "Evolution"). We also saw this with the formation of the Congo River and the separation of a group of chimps who became isolated from the main group, and who, in time, evolved to be what we now know as bonobos, who do not interbreed with chimps. So speciation occurs in localized areas, in punctuated equilibrium, which would make it far less likely that we would find evidence of the gradual changes predicted by Darwin.

Conceiving punctuated equilibrium as a descriptor of intelligent adaptation in nature for small and localized groups may make sense. Rather than evolution through small hereditary differences randomly selected over time, evolutionary variations occur by means of an intelligent, adaptive emergence spawned by a given environment along with a self-organizing dynamic, the life force that drives the universe and drives the local ecosystem as well.

Brian Goodwin, a prominent developer of theoretical biology, which uses mathematics and physics to understand biological processes, and author of *How the Leopard Changes Its Spots,* says: "New types of organism appear on the evolutionary scene, persist for various periods of it, and then become extinct. So Darwin's assumption that the tree of life is a consequence of the gradual accumulation of small hereditary differences appears to be without significant support. Some other process is responsible for the emergent properties of life...Clearly something is missing from biology" (Goodwin, 1994, p. x).

SEEING EVOLUTION AS EMERGENCE

"There seems to be no substance to the view that human language is simply a more complex instance of something to be found elsewhere in the animal world. This poses a problem for the biologist, since, if true, it is an example of true 'emergence'—the appearance of a qualitatively different phenomenon at a specific stage of complexity of organization.

—Noam Chomsky (Chomsky, 1972, p. 70).

"The world is no longer a causal machine—it can now be seen as a world of propensities, as an unfolding process of realizing possibilities and of unfolding new possibilities... New possibilities are created, possibilities that previously simply did not exist... The possibility space...is growing...Our world of propensities is inherently creative."

—Karl Popper (Popper, 1990, pp.17-20, as referenced in Rolston, 2010, p. 85-86).

The notion of emergence, which says evolution occurs in leaps or swells, with qualities more complex arising from simpler ones, might be that missing something. It is a notion now thought to be a likely way to understand the dynamics of biological evolution.

Kauffman states, "Emergence is…a major part of the new scientific worldview. Emergence says that, while no physical laws are violated, life in the biosphere, the evolution of the biosphere, the fullness of our human history, and our practical everyday worlds are real, are not reducible to physics, nor explicable from it, and are central to our lives" (Kauffman, 2008, p. x).

Steve McIntosh says, "…evolutionary scientists now agree that *emergence* is a ubiquitous characteristic of biological evolution, and emergence by definition signifies that there is a jump or a surge—that's something more has come from something less.

"Thus when we face the facts of evolutionary emergence, we can begin to see that the underlying assumption that evolution must *always* occur randomly through tiny steps and without the influence of any 'outside information' is not a scientific fact, but rather a commitment of faith held for the sake of the consistency of the theory," observes McIntosh (McIntosh, 2012, p. 35).

Emergence suggest an intelligent, intentional process in the universe, which, ironically, is unconsciously validated in the language of materialism. Narby notes the intentionality imbedded in the language of materialist biologists and wonders aloud about the implicit contradiction in the materialist interpretation: "Shamans say the correct way to talk about spirits is in metaphors. Biologists confirm this notion by using a precise array of anthropocentric and technological metaphors to describe DNA, proteins, and enzymes. DNA is a *text*, or *program*, or *data*, containing *information*, which is *read* and *transcribed* into messenger-RNAs. The latter feed into ribosomes, which are *molecular computers* that translate the *instructions* according to the genetic *code*. They build the rest of the cell's *machinery*, namely the proteins and enzymes, which are *miniaturized robots* that construct and maintain the cell. Over the course of my readings, I constantly wonder how nature can be devoid of intention if it truly corresponded to the descriptions biologists made of it…." he muses (Narby, 1998, p. 137).

Then Narby frames the essential conflict between inanimate materialism and an intelligent universe: "How could nature not be conscious if our own consciousness is produced by nature?" (Narby, 1998, p. 138). This captures the contrast in assumptions between a purely mechanical, cause and effect universe on one hand, and a vital, deliberate, and agentic intelligence operating on the other.

"The new scientific account of the operation of *emergence* challenges the reductionistic, physics-based approach," says McIntosh, *"not by positing the emergence of new physical substances per se, but by nevertheless recognizing higher levels of organization that cannot be reduced to lower level explanations"* [emphasis added] (McIntosh, 2012, p. 11).

"System science's breakthrough," says McIntosh, "came from the discovery that all evolution in the universe occurs through the processes of *self-organizing dynamic* systems that channel or metabolize energy as a method of developing and maintaining their organization" (McIntosh, 2007, p. 24). He references evolutionary biologist Peter Corning, who defines emergence in systems as: "the arising of novel, coherent structures, patterns and properties during the process of self-organization in complex structures" (Corning, 2005, as referenced in McIntosh, 2012, p. 10).

McIntosh states that: "while this definition may be technical correct, it fails to communicate the radical *novelty* and *creativity* that accompanies the appearance of an entirely new level of reality, which the most dramatic incidences of emergence represent. For example, out of the 'singularity (the theoretical state of things prior to the big bang), all at once there emerges space, time, energy, matter, and the laws of physics! [If not "rosebuds, giraffes, and humans!"] In other words, at first there was almost nothing, and then in an instant there was something—something very big. It doesn't get more radical than that" (McIntosh, 2012, p. 10).

He continues, "Today, the 'experts' on evolution generally recognized by mainstream academia and the corporate media are a closely-knit group of scientists known as 'Neo-Darwinists.' Neo-Darwinists are firmly committed to the metaphysical principle that, like physics, biological evolution is essentially a mechanistic process that can be completely explained using reductionistic methods [such as "natural selection" and "genetic mutation" processes]. For example, Neo-Darwinists hold that macroevolution (major transitions in species or taxa) is to be understood entirely by the processes involved in microevolution (accumulation of variations in populations) (McIntosh, 2012, p. 33).[142]

Microbiologist James Shapiro, an expert in bacterial genetics, is among the growing numbers who argue that the standing view of Neo-Darwinian evolution is inadequate. His data shows that evolutionary change is an active cell process capable of making rapid large changes. He says in *Evolution: A View from the 21ˢᵗ Century* that "The capacity of living organisms to alter their own heredity is undeniable. Our current ideas about evolution have to incorporate this basic fact of life" (Shapiro, 2011, p. 2). *Innovation,* not selection, is the critical issue in evolution, he argues, because innovation creates novelty, and selection has nothing to act on. And he writes in *Origins: Skeptics Guide to the Creation of Life on Earth:* "In fact, there are no detailed Darwinian accounts for the evolution of any fundamental biochemical or cellular system, only a variety of wishful speculations. It is remarkable that Darwinism is accepted as a satisfactory explanation for such a vast subject—evolution—with so little rigorous examination of how well it's basic theses work in illuminating specific instances of biological adaptation or diversity" (Shapiro, 1996, p. 94).

"It turns out that biological evolution by Darwin's heritable variation and natural selection cannot be 'reduced' to physics alone..." says Kauffman. "For the reductionist, only particles in motion are

---

142 Neo-Darwinists go as far as claiming that evolution may not progress. McIntosh comments, "One of the fundamental questions raised by science's investigation of evolution is whether it progresses. The Neo-Darwinians have generally eschewed the idea of 'higher' and 'lower' forms of evolution, claiming that such ranking involves a value judgment that has no part in the scientific discourse. The very idea of progress of evolution has been attacked by a number of prominent scientists, including biologist Stephen Jay Gould, who argues that the only legitimate criterion for progress is the effectiveness of an organism's adaptation to its local environment and by this criterion, the most progressive organism are bacteria because they have shown themselves to be the most successful at adaptation. Following this reasoning, many in the scientific community claim that a major triumph of the modern evolutionary synthesis has been the removal of outdated ideas about direction, progress, or increasing value in natural evolution. Advocates of scientism argue that the science of evolution has shown how undirected forces, blind and unintelligent are the *sole* causes that have produced the order we observe in biological systems, and that this proves that evolution is not progressive. Thus, while they maintain that science cannot make value judgments about higher and lower, Neo-Darwinians are perfectly willing to claim the status of 'science' for their value judgment that there is no value in evolution" (McIntosh, 2007, pp. 274-275).

ontologically real entities. Everything else is to be explained by different complexities of particles in motion, hence are not real in their own ontological right. But organisms, whose evolution of structures and processes, such as the human heart, cannot be deduced from physics, have causal powers of their own, and therefore are real *emergent* [emphasis added] entities in the universe. So, too, are the biosphere, the human economy, human culture, human action" (Kauffman, 2008, p. 3). Intelligence is manifest in all of nature—an agentic capacity to know!

All of these things make sense and are part of the inclusive and unitary Consciousness Paradigm of quantum realism, of the primacy of consciousness. Reductionist science seems to have it backwards. It is not out of inert matter that life and mind arise. Rather it is out of the intelligence of the quantum mind that matter, life, and self-reflective consciousness arise—emerge—in an ever-deepening spiral of evolution.

McIntosh writes about the status of Neo-Darwinian evolution that, "In the last decade or so, some biologists have begun to question the strict doctrines of Neo-Darwinism, which insist that random changes in genes selected by a changing environment provides a sufficient explanation for practically all forms of biological evolution. These more open-minded biologists have begun to see that evolution is a multi-level process of co-evolution that occurs between genes and ecosystems, [recognizing] additional evolutionary mechanisms besides natural selection, such as group selection, symbiosis, and horizontal gene transfer…Discoveries in the field of [developmental evolutionary biology—evo-devo] have shown how vast differences in organismal forms can arise from the same set of genes through variation in the timing or pattern of gene expression, rather than through changes in the genes themselves" (McIntosh, 2012, p. 37).

Genes and genetic variations, ecosystems, natural selection, group selection, symbiosis, horizontal gene transfer, *and* the panoply of flora and fauna and their intelligent and ever-creative and innovative dance, which, right now, are evolving, exploring, expanding; consciously communicating and cooperating—all are emerging as a vital, interconnected, conscious, intelligent, intentional unity!

A final challenge that can be levied against the assumption that the mechanisms of natural selection and genetic variation ought to be viewed as the primary forces and as unarguable realities in the theory of evolution has to do with whether the theory can be refuted. Philosopher Karl Popper, who is generally regarded as one of the greatest philosophers of science of the twentieth century, developed what is called the "falsifiability criterion," which states that if a theory cannot be shown to be false, it is considered to be valid. He "argued that one could never prove a scientific theory to be correct, because only an infinite number of confirming results would constitute definitive proof...

"Popper proposed instead to test theories in ways that seek to contradict, or falsify, them; the absence of [contrary] evidence thereby becomes proof of the theory's validity" (Popper, 1974, p. 171). Popper says about Darwinian evolutionary theory: "I have come to the conclusion that Darwinism is not a testable scientific theory but a *metaphysical research programme,* a possible framework for testable scientific theories....It is metaphysical because it is not testable...Darwinism does not really predict the evolution of variety. It therefore cannot really *explain* it. At best, it can predict the evolution of variety under 'favorable' conditions" (Popper, 1976, p. 168).

So Popper is saying that the dogma of our current evolutionary theory, which declares that natural selection and chance genetic variation are the cornerstone mechanisms of evolution, is a philosophical viewpoint, not uncontested scientific theory. Reiterating this perspective, Pierre Laszlo, French chemist and Princeton University professor, also an expert in nuclear magnetic resonance, stated: "The origin of life is more a question of metaphysics than a scientific problem. The experimental facts gleaned by different well-established authors allow only for scenarios, in an unlimited number, all of which are fictive" (Laszlo, 1997, p. 220).

What are the implications of Popper's clarification that contemporary evolutionary theory is philosophy, not science, (that is, a preferred article of faith rather than proven empirical fact)? As with the now patently slapdash and ill-founded belief that the world is purely material and mechanistic, which we have seen remains prevalent in our culture today, such a stance guarantees false conclusions. Science's difficulty

A Universe Full of Magical Things

with faith-based, mythical religion is blindly playing out in materialistic science's own reductionist, faith-based myths that curiously still hold sway in many halls of science today, as we've seen. The net result is a loss for us, wherein our tunnel vision of the world limits our understanding, potential, and intellectual and spiritual satisfaction, hinders our ability to advance our species and ecosystem, and deprives us of the greater truths and perspectives we might otherwise reach.

Narby notes that, "Biology is currently divided between a majority who consider the theory of natural selection to be true and established as fact, and a minority who question it." He believes that, in his view, "a new biological paradigm is still a long way off" (Narby, 1998, p. 145). However, in this book I've articulated a view of evolution as emergence in the context of a Consciousness Paradigm: a paradigm of a reality that is vital and yet virtual; a reality that is in the process of being enunciated by thoughtful and expert people, like the many mentioned here, that points to a new—and powerful—paradigm, one that is not "still a long way off," but one that is *already* being formulated and incorporated.

If we are willing to challenge the dominant materialist articles of faith, and consider instead that consciousness drives our reality, a new, richer and more useful scientific worldview doesn't seem so far away. If intelligence pervades nature as we have seen, if the capacity to know exists to some degree in everything, down to the electron, and if the upshot is agency, then it is easy to apprehend that evolution—via a range of dynamics, including innovation, genetic variations, the inborn intelligence of the world's flora and fauna, ecosystems, energic forces, natural selection, group selection, symbiosis, horizontal gene transfer, and so on—is driven by and *emerges out of intelligent decisions* at both the macro and micro level, and, as well, holistically, throughout nature.

Natural selection and genetic mutation are undoubtedly two clear mechanisms by which species evolve. The issue isn't whether these mechanisms apply, it seems, (albeit in a far more restricted manner than mainstream science attests), but what force or process catalyzes selection or mutation actions—mechanistic programming, as "a consequence of the gradual accumulation of small hereditary differences," with chance duplication errors being the only explanation for novelty in evolution? Or an intelligent, goal directed, informed, agentic, and conscious process—that nature chooses and transmutes in any given environment using innovation and creative adaptations that are aligned with unambiguous goals and intentions? If the world appeared as a solely mechanistic, pinball-like dynamic, perhaps the traditional perspective would resonate, but if we simply factor in the intelligence in nature's flora and fauna we have only briefly surveyed above, we see that the world is far more nuanced and agentic than that. Evolutionary theory incorporates selection and variance, but in an intelligent context—by a *capacity to know* that is part of all of nature, utilizing discrete data which *informs* selection and cultivates innovative, emergent options from the copious and wide-ranging environmental resources available.

Again, Lipton: "Evolution is neither random nor pre-determined, but rather, an intelligent dance between organism and environment." Evolution emerges out of intelligent innovation and informed selection, a process, that is not as much "survival of the fittest"—resulting from "the will to survive inherent in the *fittest*," observes Lipton, but rather more like the perspective of Alfred Russel Wallace, who contemporaneously with Darwin, described the dynamic of evolution in nature, and "who recognized that evolution was driven by the elimination of the *weakest.*" Then Lipton clarifies: "In a Wallacean world, we would *improve* not to be the weakest, but in a Darwinian world, we *struggle* to acquire the status of being the best" (Lipton & Bhaerman, 2009, p. 116).

In the final analysis, evolution is an event, an action, an emergence, information being processed in consciousness, a participatory and agentic quantum unfolding, significantly driven by, as Shapiro noted, "the capacity of living organisms to alter their own heredity."

THE LIVING COSMOS

The notion that the entire cosmos is an intelligent living entity is a shocking statement—especially since we have been so conditioned to understand the cosmos, the universe, the galaxies, the solar system, and the planets as so much characterless and lifeless matter. But as

Sahtouris, Lipton, Elgin, Narby, Kauffman, de Quincey, and many others have explained to us, what we think about as the essence of life incorporates several key ingredients which, when we examine them, broaden and deepen the definition of life, and greatly expand our big picture view of how the cosmos and everything in it functions as one interconnected, interdependent, intelligent, and alive process.

"Our universe, our cosmos, has always been a dance of interactions among large and small moving patterns, each contributing to the other's formation. It...evolved as a dance between great and small," observes Sahtouris (Sahtouris, 2000, p. 19).

Sahtouris for many years has been a truly significant yet somewhat unheralded visionary on the subject of the aliveness of the Earth as well as the universe as a whole. In the book *Mind Before Matter,* she bluntly states, "Instead of projecting a universe of mechanism without inventor... I propose that there is reason to see the whole universe as alive, self-organizing endless fractal levels of living complexity as reflexive systems learning to play with possibilities in the intelligent co-creation of complex evolving systems... I propose that it is possible to create a scientific model of a living universe...that...is not only scientifically justified but can lead to the wisdom required to build a better human life on and for our planet Earth.

"...We can construct a new scientific model...that takes into account the entire gamut of human experience and recognizes the cosmos as fundamentally conscious and alive ... a holarchic, evolving, intelligent process intrinsic to the cosmos itself (Sahtouris, 2007, Chapter 7).

What has become clear, if one can only set aside the pre-existing assumptions of the traditional/materialistic/reductionistic/deterministic and highly politicized present state of affairs in the orthodoxy of modern science, is that, large and small, the universe and all the seemingly separate features within it are one—a single, interconnected, well-orchestrated, highly complex, intelligent, deliberate, agentic, emerging/evolving, and self-reflectively conscious aliveness. If we have life, mind, intelligence, consciousness, and self-reflective awareness in "the world," these were not capacities that did not exist potentially, and then suddenly existed, any more than there is an empty oven, and then, without effort or explanation, that delicious, warm, tantalizing cherry pie sits, eager to be devoured. Such a fanciful imagining is not any better than the creation myths of the world's religions, and such a reductionist myth as the science myth of "matter first" doesn't even merit points for creativity, imagination, illumination, consistency, or believability.

The Materialist Paradigm describes evolution as unthinking, mechanistic, arbitrary, random, and as resulting from mistakes ("mutations" and "variations"). From one perspective, it's unequivocally absurd: a non-living universe defaults to random and transmuted "selections" (Really? In a cause-and-effect world, *selected* by what or whom?), to shape a living (though not-really-alive?) world, a world that has nevertheless become mindful and conscious and self-reflective, somehow! Does not it make more sense, look more consistent, feel more comfortable to align with, and seem richer and more satisfying to face the obvious? The world is not, after all, physical and inert. Rather it is aware and alive and it chooses. The brilliant, creative mind of Consciousness all along the way has made choices— intelligent, deliberate choices— ever-emerging towards a goal of a somehow better world.

Sahtouris writes, "Recent discoveries in physics strongly suggest that the nature of the universe was from the beginning such that it would come alive however and wherever possible. (Sahtouris, 2000, p. 78). How far does that aliveness, that awareness, that capacity to know, reach? How far does consciousness reach?

Chapter Ten

⚊⚊⚊⚊⚊

NATIVE HUMAN
CONSCIOUSNESS—
THE EXPONENTIAL EVOLUTION
OF CONSCIOUSNESS

THE LANDSCAPE OF CONSCIOUSNESS

"It's all about consciousness—literally and truly, all of it."

—Eben Alexander (Alexander, 2017, p. 70).

*"I arrived at the conclusion that 'Consciousness' (with a
capital 'C') is 'what is.' Consciousness, I came to accept,
is Being, and therefore all 'beings' are conscious. They do
not possess consciousness as something added on.*
*"I finally adopted the position that all things, organic and inorganic,
are conscious because they are made of Consciousness, but not all
things are alive. The universe, including us, is conscious, not as
a quality but as essence—and that part of the universe that is
us (biological beings) is conscious and alive. Thus rocks, which*

clearly are not alive, are nevertheless conscious—echoing de
Quincey's position that consciousness goes all the way down."

—Harville Hendrix, in *Radical Knowing*
(de Quincey, 2005, pp. xix-xx).[143]

De Quincey noted that esteemed physicist David Bohm "believed that a complete theory of the cosmos must take account of consciousness. This must be so because clearly consciousness is an undeniable reality in the universe. Without it we would know absolutely *nothing*. All knowledge of the universe—in science, in philosophy, in art, in religion, in mysticism, in ordinary daily life—exists only because consciousness is present to experience and register the existence of what we [relate to as] the physical world. A complete cosmology, then, must include the *knower* as well as what is known. We need a cosmological story that has a place for the *storyteller*" (de Quincy, 2005, p. 163).

We've seen that the universe is a living interconnected oneness. Within it exists intelligence, capacity to know, intentionality, agency, authorship, and a variety of other features we've identified. We are not saying that all of the universe is self-reflectively conscious, as are humans, but that consciousness exists in all of the universe on a continuum, the insight Hendrix expressed above.

A good means of refining our understanding of the range of features implicit in consciousness is this description from de Quincey:

"What kinds of attributes characterize consciousness and distinguish it from nonconscious entities? If some creature has consciousness it will be capable of the following (a nonconscious entity will lack all of these):

1. Sentience/feeling (capacity to experience)
2. Subjectivity (capacity for experienced interiority; for having a unique and privileged point of view)
3. Knowledge (capacity for knowing anything)
4. Intentionality (ability to point at, or be about something else)

---

143 Harville Hendrix is a contemporary psychologist, therapist, educator, and pastoral counselor. Along with his wife, Helen LaKelly Hunt, he has developed a form of marriage and relationship therapy known as Imago Relationship Therapy.

5. Choice (capacity to create 'first cause'; to move itself internally)
6. Self-agency (capacity to orient and/or move itself externally)
7. Purpose (capacity to aim at a goal)
8. Meaning (capacity to be intrinsically 'for-itself')
9. Value (capacity for intrinsic worth)

"To be a creature with consciousness means to be a creature that possesses any or all of these characteristics. Notice that none of these characteristics is amenable to mechanistic reduction, measurement, or objective description.

"Consciousness encompasses all experience—all sentience, all feeling, subjectivity, all interiority—the 'within' of things, as Teilhard de Chardin poetically expressed it; the 'whatitfeelslike from within' as Thomas Nagel put it. It is whatever endows an entity, being, or thing with the ability to 'prehend,' to take into itself the essence of what it experiences" (de Quincey, 2002, pp. 63-64).

By these criteria, every so called "material entity" in the universe is conscious. However, because we don't experience a rock or a chair as having any of the nine qualities above, it is easy and conventional to conclude they are not conscious. Yet the double slit experiment, so foundational to quantum theory, demonstrates—as Dyson and others have pointed out—that even a subatomic particle makes a choice: "It appears that mind, as manifested by the capacity to make choices, is to some extent inherent in every electron" (Frankenberry, 2008, p. 372).

Additionally, we've described our world as a virtual and non-physical reality, an illusory reality. From this perspective, reality is an "undivided whole," as Bohm observed, a "holographic movement." There is no separation. All that we experience as separate is indeed, one interconnected being. And if that is so, the whole and all it contains must be conscious![144]

For McIntosh, along with Teilhard de Chardin, the degree of consciousness and degree of complexity go together: "Teilhard [de

---

144 We tend to treat that which seems to act in the world as living, and that which doesn't seem to act as non-living. That's our conventional agreement, dependent on how we define "consciousness" or "life." That's how Hendrix, above, sees it. However, it's somewhat of a semantic game. Personally, I prefer to conceive that anything conscious is, by definition, alive. In part that is determined by what we consider to be conscious, which I would understand as "every naturally occurring form of universal organization." But clearly, there is nothing resembling a consensus!

Chardin] advanced his 'law of complexity-consciousness' which holds that consciousness develops in direct proportion to an organism's organizational complexity. According to de Chardin, 'complexification due to the growth of consciousness, or consciousness the outcome of complexity: experimentally the two terms are inseparable:" (Teilhard de Chardin, 2004, p. 174, as referenced in McIntosh, 2012, p.174) the greater the complexity, the greater the consciousness.

McIntosh, too, believes that "the most advanced understanding of consciousness recognizes that some primitive forms of it 'go all the way down,' that awareness pervades the entire universe and can be found at the heart of *every naturally occurring form of universal organization* [emphasis added]. Human consciousness thus shares a connection with all forms of consciousness."

De Quincey writes, "There is an ordering principle, a kind of mind, at play in nature at every level of existence—a co-creating complement to the chaos-inducing principle of entropy. This proto-psychism is consciousness in its most basic state—what influential philosopher Alfred North Whitehead[145] called 'prehension,' an elementary awareness that directs the motion of subatomic particles and atoms in their interactions.

"Intention, then, some degree of directional awareness, is active throughout the entire span of evolution," continues de Quincey, "from the simplest quantum entities to the largest superclusters of galaxies and the most complex nervous systems and brains. This view allows for the existence and evolution of the world long before the arrival of

---

145 Alfred North Whitehead, (February 15, 1861-December 30, 1947), "is best known as the defining figure of the philosophical school known as process philosophy," states Wikipedia, "which today has found application to a wide variety of disciplines, including ecology, theology, education, physics, biology, economics, and psychology, among other areas.

"In his early career Whitehead wrote primarily on mathematics, logic, and physics… Beginning in the late 1910s and early 1920s, Whitehead gradually turned his attention from mathematics to philosophy of science, and finally to metaphysics. He developed a comprehensive metaphysical system which radically departed from most of western philosophy. Whitehead argued that reality consists of processes rather than material objects, and that processes are best defined by their relations with other processes, thus rejecting the theory that reality is fundamentally constructed by bits of matter that exist independently of one another. Today Whitehead's philosophical works— particularly *Process and Reality*—are regarded as the foundational texts of process philosophy." It is noteworthy how consistent with quantum theory are Whitehead's views.

Wikipedia notes that "Whitehead's process philosophy argues that 'there is urgency in coming to see the world as a web of interrelated processes of which we are integral parts, so that all of our choices and actions have consequences for the world around us.'"

thinking beings who could believe in it or believe in anything about it. It also allows us to trace the long lineage or continuum of intentions, of innate expectations, way back to the origins of the universe…

"Intention is our creative contribution to the unforced, natural expression of these deep, embedded innate purposes common to all life, planet, universe, and cosmos," concludes de Quincey (de Quincey, 2005, pp. 59-60).

McIntosh reiterates this perspective noting, "Whitehead's conception of consciousness [is] not as something that emerges only in higher animals, but as an aspect of the cosmos that can be found inside every naturally occurring universe structure, no matter how small. Whitehead's hypothesis is that the essential 'stuff' of nature is creative, experiential events, or 'occasions of experience,' rather than simply particles of matter [which aligns elegantly with quantum theory]. Thus, according to process philosophy, all outsides have an inside. In other words, every naturally arising universe structure, from atoms, to cells, to mammals, to humans, has both subjective mode and an objective mode.[146] These internal and external aspects of reality arise together from the beginning, interact naturally and nondualistically, and pervade the universe at every level of organization. This is not to say that atoms, for example, are actually conscious in the same way humans are conscious, but that they do possess a primitive form of awareness, which Whitehead called 'prehension.' That is, every atom demonstrates rudimentary awareness in the way it interacts with other atoms. If an atom's orbital shell is empty, this state of being serves as a signal to other atoms that their electrons may enter the empty shell, and this basic form of proto-communication reveals the primordial nature of 'prehensive' awareness" (McIntosh, 2007, p. 170).

As with other emergent properties such as life and mind, de Quincey further argues that self-agency, or choice, must be an emergent

---

146 Unfortunately, here, McIntosh, who is attempting it seems to be speaking from a Consciousness Paradigm perspective, may be conflating inside and outside with subjective and objective: they are not the same thing. From a Consciousness Paradigm point of view, there is no objectivity. Inside and outside both refer to a subjective observation. He clearly states that the essential stuff of the universe is not "simply particles of matter." Yet here, as well as elsewhere, he tends at times to be speaking from a bit of a Materialist Paradigm perspective, suggesting he holds a panpsychist rather than an idealist frame of reference. To reiterate: the Consciousness Paradigm holds that, as quantum theory avers, there is no such thing as matter.

property of the universe as well: If "lower-level entities were completely devoid of choice (as mechanism assumes), how could we account for the emergence of choice *anywhere* in the hierarchy?" he asks. "How could entities lacking all capability for self-agency, free will, or choice ever give rise to entities that did have that capability? The only way we can coherently explain the existence of choice is that it must go 'all the way down' to the lowest-level microentities. Choice—the exercise of free will and self-agency—must exist to some degree at the micro-level if it exists at the macro-level. Otherwise, the emergence of choice from the utter absence of choice would require a miracle" (de Quincey, 2002, pp.25-26).

Eloquently encapsulating the same conclusion, the illustrious and prominent French philosopher, Henri Bergson, declared: "Consciousness correspond exactly to the living being's power of choice" (Bergson, 1998, p. 253).

With consciousness, comes choice, Bergson is saying. Of course, if there is no consciousness, there would be nothing making a choice, so no choice. But is consciousness merely an observing or witnessing power? Or does it influence what occurs in our world, our experience? In quantum physics and in our experience, choice is central, choice exists, choice occasions the unfolding universe.

"If 'your choices' just happen, if they are utterly random, you have no claim to agency. Similarly, if your 'choices' don't just happen, if they are not random, they must be determined, and once again you have no claim to agency. Without agency, without the power or possibility to freely initiate action, what has become of your free will?" asks de Quincey (de Quincey, 2002, p. 96). Further he considers that, "To an observer, total randomness is indistinguishable from the exercise of free choice. But to an entity with self-agency, there was a world of difference between randomness and the experience of choice" (de Quincey, 2002, p. 236).

We can look at choice, or free will, in the context of the three points of view we might take, as I presented in the introduction: A Materialistic Paradigm is deterministic, thus there is no choice. A Consciousness Paradigm (quantum realism) implies choice throughout the universe, including among the smallest and largest of what we see as physical entities, which would be so if everything is interconnected, as quantum theory says. However, in the viewpoint of the Quantum Mind Paradigm, wherein there is no time or space or matter, there is no choice, because nothing is acting on nothing: the world that manifests itself to us is an illusion. Therefore, there really is no argument about the "existence" of choice. It's a matter of how each of us subjectively "sees" the world. Whether choice exists is a matter of which perspective we take: a Materialistic Paradigm, a Consciousness Paradigm, or a Quantum Mind Paradigm.

The exercise of choice, self-agency, also implies there are options to choose between or among. "In order to make choices, minds must contain alternative possibilities co-existing at the same time. In the language of quantum physics, these possibilities are 'superposed,'" comments Rupert Sheldrake, (Sheldrake, 2012, p. 228), reaffirming the link between quantum and mind.

Then Sheldrake joins Freeman Dyson in asserting that human consciousness and subatomic consciousness are only differing gradations of the same phenomenon. Sheldrake cites Dyson: "The physicist Freeman Dyson wrote, 'The processes of human consciousness differ only in degree but not in kind from the processes of choice between quantum states which we call "chance" when made by electrons'" (Sheldrake, 2012, p. 228). Self-agency exists throughout.

So consciousness pervades our existence. Although there are "physical" structures associated with consciousness, as with humans, consciousness is not localized in any "physical" entity. Consciousness suffuses existence.

It is not a phenomenon that is sited in the brain. "Everything seems…to happen *as if* consciousness sprang from the brain, and *as if* the detail of conscious activity were modeled on that of cerebral activity," notes Bergson. "In reality, consciousness does not spring from the brain; the brain and consciousness correspond because equally they measure, one by the complexity of its structure and the other by the intensity of his awareness, the quantity of *choice* that the living being has at its disposal" (Bergson, 1998, p. 262).

McIntosh makes clear: "Although the internal and external are recognized as different phases of the same thing, that 'thing' is not merely particles of matter" (McIntosh, 2007, p. 16). He says, "Through the advances of culture, the consciousness of individuals has shown its ability to evolve away from the underlying evolution of the brain. While there is a neurological basis for all conscious activity [in humans], the mind's demonstration of its ability to evolve and grow, even while the structure of the brain remains relatively constant, makes clear that the mind takes form around structures that cannot all necessarily be found in the neurological physiology of the brain. And the evolving system of values-based worldviews that we recognize as the spiral of development is a prime example of a psychic structure that interacts with the brain which is not made of gray matter" (McIntosh, 2007, p. 242).

"Human consciousness is also unlike any other form of consciousness we know of because *human consciousness has demonstrated its ability to evolve in ways that do not depend solely on the corresponding evolution of its underlying biology*" [emphasis added] (McIntosh, 2007, pp. 11-12).

In the story of evolution, McIntosh's comment here points to a whole new component of evolution: the emergence out of biological evolution into a noetic capacity in humans that transcends and includes mere biological evolution. McIntosh heralds this momentous shift in evolution's unfolding: "The appearance of human consciousness constitutes an evolutionary breakthrough as significant as the [presumed] original emergence of life from inanimate matter—because human consciousness is uniquely self-[reflectively] conscious... With self [-reflective] awareness comes the ability to take hold of the evolutionary process itself" (McIntosh, 2007, p. 13). Biology has ceded its place as the cutting-edge laboratory for evolution's work to native human self-reflective consciousness, which, we have seen is indisputably progressing in its complexity.

Although the reductionist conception of evolution as material, random mutation and natural selection makes little or no room for a progressive arc in evolution, "Science does recognize at least one direction of evolution's progress, and this is the direction of expanding systemic complexity, or an increase over time in the number of functional parts in any given system...a direction in evolution that many scientists are willing to acknowledge. The progressive nature of increasing complexity is well articulated by Erich Jantsch, who writes, 'the evolution of the universe is the history of an unfolding of differentiated order of complexity…Complexity…emerges from the interpenetration of processes of differentiation and integration.'" (Jantsch, 1980, p. 75, as referenced in McIntosh, 2007, pp. 275-276).

McIntosh comments further that "System scientists like Jantsch have generally been more willing than Neo-Darwinian biologists to recognize evolutionary progress; in fact, some prominent system scientists have actually celebrated it" (László, 1996, p. 44, as referenced in McIntosh, 2007, p. 276).

Apparently, then, the form evolution has selected for its unrelenting push toward organization and integration, complexity and individuation is *native human consciousness.* As this realization begins to dawn on humanity in ever more intuitive and comprehensive ways, perhaps we will become more deliberate and intentional and choose to partner with evolution, to partner with the life force of the universe, and create *organizations* that enhance our environment and citizenry's experiences, that *integrate* all demographics and all ecosystems, that morph *complexity* in the directions of well-being, and that promote a deepening and expansion of *individual* personal consciousness within humanity (Stewart, 2000), all the while recognizing our interconnectedness.

Graham Harvey offers a big picture perspective for us: "We have never been separate, unique or alone and it is time to stop deluding ourselves. Human cultures are not surrounded by 'nature' or 'resources', but by 'a world full of cacophonous agencies', i.e. many other vociferous persons. We are at home and our relations are all around us. The liberatory 'good life' begins with the respectful acknowledgment of the presence of persons, human and other-than-human, who make up the community of life" (Harvey, 2005, p. 212). De Quincey aptly summarizes this sentiment, when, while pointing plainly toward humanity's evolutionary work, he notes that "the mark of a good life is the quality of our interconnectedness" (de Quincey, 2005, p. 6).

The partnering of biological evolution with native human consciousness will move us beyond genetics toward whatever we humans can value or imagine. Cognitively and technologically, we are moving at breakneck speed in our process of emergence. It is not true, however, that we have advanced to the same degree morally and ethically, and, in the end, it is only the degree to which we incorporate vales of compassion and inclusion in our evolutionary arc that we will be successful in aligning with the goals of a living, intelligent universe.

A Universe Brimming with Life and Mind

"In the infinite beauty, we're all joined in one."

— John Denver, *The Flower That Shattered the Stone.*

We have seen, beyond any doubt, that, in our perceived reality, our planet is a bountiful cornucopia of brimming life and mind, one interconnected being, ever unfolding as "an intelligent dance between organism and environment." Each separately-appearing aspect is as a human cell is to a human body: it cannot meaningfully be seen as distinct from the body or as operating independently of that body. As a whole, its features include, among others,

Unity;
Autopoietic aliveness;
A capacity to know—intelligence;
An ability to choose and to create—agency and intentionality;
Evolution;
Innovation;
Creativity;
Cooperation;
Emergence;
Complexity;
Relationship;
Consciousness;
Meaning and Purpose, and thus, Value.

These features ensue at the subatomic level, at the cellular level, and within and throughout the universe, the cosmos-at-large. How could these features not be pervasive in the cosmos, when, as we have seen via cosmology and the theories of relativity and quantum, that the cosmos does not consist of discrete entities, but is instead one entity? I trust, our guest representatives[147] have shown these things to us persuasively, describing a synchronous, consistent, and elegantly harmonious symphonic unfolding.

The ways we have been taught to understand our existence here, ways that were, in the past, the deepest wisdom our culture had to offer, must be revised and updated. We can no longer limit ourselves by aligning and identifying with an outdated perspective. We can no longer deprive ourselves of the richness, depth, purpose, meaning, and joy that this Consciousness Paradigm of our interconnected aliveness in consciousness, which we have been describing, has to bestow.

That paradigm reveals to us the intelligent aliveness and a pervasive wisdom that underlies our being. It informs us as well that, fundamentally, this story we are living, each from our own vantage point, each playing our assigned role, is *not,* finally, what is real. It is real for us. It is real to us. But it is a limited reality, a pretend reality of separate things and physical forces—a virtual reality, like a computer game we play, and will, at some point, put down.

This new paradigm underscores, as well, that there indeed is a first and final cause: quantum fluctuations, God, or Quantum Mind. There is an uncaused cause, and whatever it is, it is responsible for our being here. We cannot conceive of something from nothing. We observe it when we witness emergence—the big bang, life, mind, language, new species or new physical attributes, or, seemingly as well, the breeze. But we don't have any experience that can address how and why; we have no way to wrap our heads around an idea such as first cause. All we have is curiosity and speculation.

And yet, our virtual reality compels us forward: an inevitable and continual evolutionary unfolding and becoming and emerging of life and consciousness. Like the persistent and varied and animated bird sounds one hears in an aviary, the evolution of our cosmos and our Earth reverberates as well with tenaciousness and diversity and vibrancy. In

147 ...such as Ackerman, Bell, Bohm, Bohr, Born, Capra, Chopra, de Duve, de Quincey, Dossey, Einstein, Eddington, Einstein, Elgin, Gotswami, Greene, Heisenberg, Hoffman, Huxley, Jeans, Kauffman, Kochen, Lipton, Lovelock, Rae, Rolston, Rosenblum and Kuttner, Laszlo, Lanza, Leibnitz, McIntosh, McTaggart, Miller, Naess, Narby, Penrose, Planck, Pollan, Rees, Safina, Sahtouris, Schrödinger, Vertosick, von Braun, Walker, Ward, Whitworth, Whitehead, Whitesides, Wilber, Wohlleben, Zeh, and a bounty of others.

forms that are physical, biological, cultural, mental, emotional, social, relational—and spiritual—in venues that are internal and external, and in processes that evolve and emerge, the living universe continues, day by day, to reveal to us—to US!— more and more of its exquisite, indescribable brilliance, beauty, and mystery. There is nothing inert or lifeless about it.

BEYOND BIOLOGY: THE INTERIORS OF CONSCIOUSNESS

"So I embrace what isn't known,
And accept what we are, we can't believe.
Cuz everything you need to be, you already are.
And everything you need to know, is in your heart.
It's in the pouring rain, it's in the season's change, it's in the flowing stream,
It's in the barreling waves, it's in the highs and lows, it's intermingled toes,
It's in your lover's eyes, it's in the moon's full tides.

—Julian Yeats, ("Everything You Need to Know," *Higher Vibe* 2018).

"Not only is matter and life evolving in the exterior realm, but
evolution is also fully active in the internal realms as well.

–Steve McIntosh, *Integral Consciousness* (McIntosh, 2007, p. 217).

Evolution occurs in a variety of areas: geology, cosmology, anthropology, and biology, in medicine and technology, and cognition, to name a few. And in consciousness. And evolution's interactions among and within these and various other areas result in an expansive and interacting complexification beyond that of any one arena. The movements, processes, interactions, direction, and agentic intelligence of evolution illuminate an elaborately intertwined and spectacularly choreographed dance of nature in all its forms.

With the onset of the evolutionary process of biology, which arose at the big bang—at the very beginning—consciousness was able to express itself specifically among the parade of life forms and among each of the naturally occurring entities emerging in the universe. In humans, biology provided a structure wherein consciousness could evolve beyond the gradations of existential sentience common to all beings and all life forms, into self-reflective, intentional consciousness—individual personal consciousness—resulting in language, culture, civilization, technology, and an unarguable inner subjective realm that allows us to participate, to observe, to experience, to share, to think, to feel, to imagine, and to communicate and act together. Quantum dynamics consistently demonstrates that there is no objective reality: it is individual and subjective consciousness, this individual personal consciousness, which defines our core reality.

Steve McIntosh makes the powerful point that evolution involves this personal consciousness, which, as we perceive it, itself is evolving, *and which evolves somewhat independently from biological evolution*—as we will discuss shortly. As evolution progresses and complexifies, it displays increasing consciousness, increasing degrees of awareness: "When we include the development of mind or consciousness in the evolutionary picture," he says, "there can really be no doubt that evolution has progressed—that it has produced a sequential series of organisms that exhibit increasing degrees of awareness." McIntosh observes that only by *not* considering consciousness *at all* can it be argued that human evolution is no more advanced than amoebas: "It is only by leaving consciousness completely out of the picture that Neo-Darwinist philosophers can credibly avoid acknowledging that humans are 'higher' than amoebas" (McIntosh, 2007, p. 282).

Expanding on this, he tells us that "the evolutionary novelty of humans was not really a biological breakthrough. In fact, the biological differences between early humans and their immediate animal ancestors were barely noticeable." Then, he notes insightfully and importantly: "The evolutionary leap constituted by the appearance of humans was *internal* [emphasis added]—it came about through the advent of a dramatic new type of self-consciousness. This self-awareness, this consciousness of consciousness itself, appears only in humans. It is the emergence of this new self-reflecting ability in humans that marks the real beginning of the developmental domain of cultural evolution" (McIntosh, 2007, pp. 12-13).

Biological evolution continues in its deliberate trajectory, but now, with the emergence of the phenomenon of self-reflective consciousness, *native human consciousness,* mind enters into an evolutionary unfolding that is mystifying in its quickness, its intricacy, its capacities and potential, and in its very emergence. McIntosh says, "Consciousness is generally ignored by the scientists who study evolution for two reasons: because of science's inability to explain subjective phenomena, and because the inclusion of consciousness in any estimate of evolution reveals unmistakable evolutionary progress." And he reiterates, "In fact, it is only by ignoring or denying the evolutionary significance of the appearance of human [self-reflective] consciousness that scientists have been able to even question whether evolution is progressive or not" ((McIntosh, 2007, p. 282).

The advances in civilization and culture clearly evident in humanity's historical record, although they have been hosted by biological processes, have established a new, somewhat independent and transcendent arena of evolution involving native human consciousness. Biologically, humanity continues its developmental progression, partnering with other co-evolving processes to advance a mental, or noetic, realm propelling planetary change, development, and awareness at unimagined speeds, especially given the short time humans have wandered the planet.

"The evolutionary significance of human consciousness is clearly demonstrated by the now-obvious fact of global human culture," McIntosh points out. "Development and the complexity of human cultural structures is undeniable. Unlike the previous evolutionary breakthrough seen in the appearance of life, the appearance of human culture is accompanied by new methods of development and the new pace of progress. Just as life evolves much faster than inanimate matter, human consciousness and culture evolve much faster than life. However, even though the emergence of human consciousness and culture constitute a new domain of evolutionary progress, methods, habits, laws of evolution still apply" (McIntosh, 2007, p. 14).

He says that "the evolution of human consciousness actually occurs in a distinct 'domain of evolution' apart from biology… The inside of an individual (her consciousness) can evolve and develop without the outside of that individual (her biology) evolving much at all… Before the appearance of humans, an organism's inside mind and outside brain evolved together in lockstep. From the beginning, consciousness was present in rudimentary ways throughout the universe. For life to become appreciably smarter, it had to evolve biologically. But with the advent of humans, the internal domain of consciousness is partially liberated from its biological constraints and is able to embark on the path of a whole new type of mental, emotional, and spiritual evolution. However, the essence of this development is *within* consciousness and culture; this occurring in the domain that is best described as the *internal universe*" (McIntosh, 2007, p. 16).

The above statement proclaims something altogether new in the process of evolution in the universe. Mind, in humanity, evolves somewhat independently from biology, and is not bound to the pace of biological growth and evolution. The internal domain of self-reflective consciousness, it turns out, is an unarguable, distinctive, and emerging phenomenon, not previously observed in evolution. And it is the most potent illustration of evolution in our universe yet—because of its capacity to know, to know that we know, to communicate and co-create, to witness and participate, to ponder and imagine, to create, and to deliberately author new, original, distinctive, cumulative, and visionary inventions in the realms of systems, communication, society, relationship, culture, civilization, and mind, among many others.

All of this emerges *from the internal world* of individual and personal consciousness, subjective experience as well as intersubjective shared knowing. This internal consciousness, through the phenomenon of observation, participates in creating what we call "reality," again affirming Hoffman's assertion that "physical objects are not fundamental," and Lanza's biocentric argument that the internal universe is all that is.

Upon thoughtful examination we see that, *indeed subjective experience and intersubjective experience are, in fact, all that is real*—everything else we know is but an "as if," or virtual representation, like a computer icon or TV show or movie or computer game, or a sensory, four- dimensional holodeck program.

What we experience as reality exists only in *minds*. What any one of us experiences as real exists only in the mind. *Our* reality is a representation, perceived subjectively. The only reasons, it seems, we respond so incredulously when we are asked to consider that the material world itself may be virtual, is because we have little experience with matter or tactile or sensory experience being only a representation of something else, and we have little familiarity with multisensory, palpable, and three and four dimensional *symbols*. Star Trek's holodeck, it seems, may be the closest representation.

We accept how this can be so on Star Trek's holodecks: the crew interacts with virtual entities that feel material, that have a "hardness"—as well as an "aliveness," even "personality." Quantum says that, in effect, the holodeck world is the same as our experienced world, our virtual world. All that is "real" on the holodeck is the subjective experience of the crew member who participates in the virtual fantasy. The only thing that is different about a holodeck reality and ours is that the crew member *knows and accepts* that what is being lived is a virtual story.

Within this representation of what is real, within this tangible and yet symbolic, virtual experience of consciousness that is, simply, the human story, biology continues to intelligently evolve in a deliberate and brilliant, yet relatively gradual unfolding. But as McIntosh has shown, biology has passed the evolutionary baton to self-reflective consciousness, the native human consciousness it birthed. Now, a native, exclusively human type of consciousness has superseded biology, as Rolston observed. No longer is native human consciousness, (self-reflective consciousness), dependent entirely on biological evolution to spur its growth. Now, as McIntosh tells us, native human consciousness is capable in itself of evolving, reflecting, changing, growing, creating, and doing so in a deliberate and visionary manner, and to a degree *independently of biological evolution.*

Native Human Consciousness

This most recent emergent unfolding is the one that allows us to see. And to know. And to know that we know. And to communicate with each other what we know. And to wonder and explore and discover. And create. And to document it. It is a potent, almost inconceivable gift that, as far as we know, has been bequeathed to humanity alone. It is the gift of consciousness-beyond-awareness, what is called self-reflective consciousness. It is the one feature we know of in our cosmos that truly differentiates humanity from all else: it is our native human consciousness, and it is a gift that calls humanity to a level of accountability and responsibility far exceeding all else.

We are mere children in our efforts to understand and utilize this potent genius. We are, ironically, barely aware, as a species, of this subtle capability we possess and what might be done with it. So far, we have largely used this wonderful talent to accrue personal advantage at the expense of others, at the expense of our resources, and at the expense of what might be possible. But that is illustrative of the point. Consciousness, and our consciousness, are part of the evolution of the cosmos and the planet. We are not a done deal.

With this unique skill, we can, for the first time, *participate with and co-create with the universe itself.* We can be part of charting where consciousness itself can grow to. We can be a part of how our self-reflective consciousness emerges and we can be a part of authoring how we contribute to the evolutionary unfolding of the universe itself. We are not alone in our participation with how the physical world evolves—every rock and galaxy and everything in between is part of that as well. And we are not alone in our biological participation. We are not alone in our emergence into life. Nor consciousness. But we, alone, are the agents who will participate in shaping the evolution of self-reflective consciousness. We will say what we can evolve into and what can emerge from it.

It appears that consciousness has representation in nature from the extremely simple to the impenetrably complex, and there is no reason to conclude that the process will not continue to evolve to even more nuanced and dumbfounding heights and depths. Dyson's muse about awareness of some sort existing even in the electron goes to the notion that if something manifests in the universe it can be found throughout the universe. Conway's and Kochen's free will theorem,[148] says the same thing.

If our being is, as quantum theory asserts, one interconnected being, then consciousness, it would seem, has a presence in some form throughout. That seems to be true as well for life, as we now have

---

148 "If any part of the universe has free will, it all does, but if any part doesn't, then none of it does—so either nothing is conscious or everything is."

found and as we have noted: organic molecules were existing bountifully in the early gases and chemicals erupting from the big bang in a kind of interstellar nursery. So life is not to be found only in our latter-day Earthlife examples, but throughout the universe, unfolding as well, we can imagine, in lesser and more mature forms elsewhere, as on Earth. As Sheldrake noted, "Since the universe itself is evolving and developing, the mind or consciousness of the universe must be evolving and developing too" (Sheldrake, 2012, p. 338).

A conscious universe throughout does not mean we should expect that consciousness exists *in fullness* everywhere, any more than we expect that every atom of a redwood be alive. Aliveness is carried in the structure and arises through the interactions of what is living and what has or will live within the ecosystem that contains it.[149] Within early stellar explosions exist the structure and loam—and the capacity to know—for life to emerge.

Primitive versions of life possessed, we can assume now, a primitive awareness: a self-knowing of existence, and a skill-set of responses designed to self-organize, and to nourish and reproduce. Then single-celled beings over time collaborated and communicated together, toward a common aim, as Lipton described, eventually setting the stage for multicellular organisms. Each step of the way, consciousness increased in complexity and capacity, such as in establishing an ability to communicate via signal molecules—a capacity to know in common with one's fellow citizens—and, in multicellular organisms, the beginnings of specialized cells. The more complex an environment and a life entity would become, the more diverse and wide-ranging skill-sets and capacity to know would be required for its continued thriving. A bacterium requires a different set of tools than does a lion, and each a unique and specific skill-set and knowledge; and each requires an enriched and unique awareness, or consciousness: a necessary resource to meet any challenges.

Many animals, as we have explored, have grown in consciousness well beyond instinctive knowing, and choose, assess, make decisions, communicate with their own and other species, exhibit emotion and connection, cooperate and adapt, use tools, and even know

---

149 As de Quincey amplifies below.

telepathically (as in the case of the killer whale). All animals and plants, we have seen, acquire the consciousness to survive and develop to the degree their continued life requires. Consciousness is a constant, common, and pervasive, albeit variable, feature of the entire plant and animal kingdoms, it seems.

And with humans, as we have seen, consciousness advances beyond what we have observed in the animal and plant kingdoms, to incorporate what is called self-reflective consciousness, or, the ability to know coupled with the ability to know that we know (and to make choices based in that knowledge). That added capacity has allowed us to create civilizations of grand design and impressive order and complexity, and has allowed us to sate our curiosity about so much of what our existence here is and means. We can articulate through language the things we have witnessed, we can calculate and hypothesize through our sciences, and we can document our discoveries, our ideas, and our philosophies and romances through writing, so that the wisdom and conclusions and perspectives of our and previous generations and societies will not be lost, and can be more deeply incorporated going forward.

But human evolution in the realm of self-reflective consciousness is not homogeneous: just as there are gradations in the animal world, there are gradients in the human world. This is a very hot metaphysical topic, I should warn, because humans typically believe in their own conscious bias as most viable, as McIntosh illustrates: "Traditional consciousness identifies the need to reduce lawless violence and evil in the world, yet creates oppression. Modernist consciousness identifies opportunities for development and discovery, yet it creates gross inequalities. And postmodern consciousness identifies the need to honor and include everyone, it also creates blindness to comparative excellence" (McIntosh, 2007, p. 75).

Traditionalists are loyal to conventional structures and values and norms, Modernists are philosophically self-centric, valuing reason, science, and technology; postmodernists value multiculturalism and environmentalism, and claim all points of view have equivalent worth. Post-postmodernists, few in numbers as they may be, and mindful of the interconnectedness of it all, might say that there are intrinsic values shared by all humanity that supersede the provincial instrumental values of any given group.

Among those intrinsic values, most would agree, are that all humans have rights to life, liberty, respect, happiness, equality, and choice, among others, despite whatever capabilities they may have, life conditions they may be experiencing, or ecosystem (environment) they may be a part of. Intrinsically, we are of equal value in these ways. So claiming that there are deeper and shallower levels—or less mature vs. more so—in human self-reflective consciousness should not be understood to be a sanction for racism, misogyny, gender or other bias, ageism, insularity, or any other kind of –ism or devaluation or discrimination of individuals or groups.

Need, experience, opportunity, resources, and in general, life conditions, are among the variables that might catapult one individual or group toward one specific set of values vs. others, and toward more inclusive arenas of consciousness vs. less so; life conditions that are on the face of it capriciously doled out by our universe. Subjectively, my instrumental values count more to me, but in a perspective that acknowledges our interconnectedness, everyone is equally *intrinsically* valuable irrespective of the narrow instrumental scope of our provincial concerns. A three-year-old has as much intrinsic value as a thirty-year-old, though there be a vast difference in the realization of self-reflective consciousness for each.

The point is that self-reflective consciousness is an evolving and emerging process in humans, and one size does not fit all. Research has shown that human development proceeds through sequential stages, that cannot be skipped, and this is so in a variety of human growth areas—physical, emotional, moral, values, cognitive, social, relational areas as well as, for example, mathematical or musical, kinesthetic or artistic areas, or spiritual areas. Ken Wilber has delineated these areas as "lines of development," and humans vary greatly in their capabilities within these lines. Birds and fish and mammals and trees each possess skill-sets germane to their species and environment, which also vary greatly, and each has the appropriate awareness of how to employ flying skills as opposed to swimming skills or hunting skills or photosynthesis skills. Every species, and every human, has their unique subjective consciousness without, necessarily, a great deal of overlap. We're all skilled,

A Universe Full of Magical Things

but differently. And we're all conscious, but differently. Yet, being one, it is certainly in our best interest to value all parts of the whole, and work together.

John Stewart's hopeful perspective is that "the direction of evolution is towards increasing cooperation between living organisms. As evolution proceeds, living things will increasingly coordinate their actions for the benefit of the group rather than acting on their own individual interests. Cooperators shall inherit the earth, and eventually the universe" (Stewart, 2000, p. 203).

And physicist Freeman Dyson is equally optimistic: "I have found a universe growing without limit in richness and complexity, a universe of life surviving forever and making itself known to its neighbors across unimaginable gulfs of space and time. Whether the details of my calculations turn out to be correct or not, there are good scientific reasons for taking seriously the possibility that life and intelligence can succeed in molding this universe of ours to their own purposes" (Dyson, 1988, p. 117).

THE EVOLUTION OF HUMAN CONSCIOUSNESS AND VALUES

McIntosh, in *Evolution's Purpose,* notes the insight of Clare Graves that one of the important ways that human consciousness continues to evolve is predicated on both our life conditions and our identification with a given set of values. Evolution continues to occur in the arena of consciousness, which is driven significantly, he observes, by our values orientation. He says, "Values attract evolutionary development through their influence on consciousness…The evolution of consciousness and culture occurs as a result of both the *pull* of our values and the *push* of unsatisfactory life conditions."

He further observes, "Interior consciousness almost always evolves with exterior circumstances…Wherever we find the interior evolution of consciousness, we also find a corresponding evolution in the complexity of the exterior structures associated with such interior development" (McIntosh, 2012, p. 203).

A very useful illustration of the development of human values is the rich pool of research data that was obtained in the late 1950s and in the 1960s, and continuing, and which was designed by Clare Graves, professor of psychology and originator of a theory of adult human development. He initially named his interpretation of this data "The Emergent Cyclical Levels of Existence Theory" (ECLET).[150] His work was continued beyond his death in 1986, by Don Beck and Chris Cowan, under the rubric "Spiral Dynamics."

Spiral Dynamics is an extremely helpful description of how our worldviews can shift based in what we value, (in the context of our life conditions), and, as well, can indicate to us how these worldviews might further evolve in the future. Clare Graves began his research in 1951, interviewing thousands of people, and, in time, interviews were performed by hundreds of administrators and thousands of individuals in a wide swath of countries and cultures.

He did not have a theory which he then tried to do research on in order to validate; rather, he let the data speak for itself. As Don Beck says, Graves "did not design these (systems), he uncovered them, which means they existed all along, but not detected by us at this stage." This description of how humans think about things and how they function as they move through life's stages points out that not only how our worldview and our values, (that is, our conscious awareness), may evolve, but do so in predictable, describable ways.

Spiral Dynamics says our values evolve as our life conditions change. A concrete example is the difference a child experiences when she or he learns to walk. Suddenly, the world is a different place, because I can go to things, as opposed to watch them from afar, and I can interact more fully. My whole sense of what the world is, who I am, and who others are shifts suddenly, markedly, and forever. With this new view, my values change. Exploring objects by being able to hold them loses its luster: I am much more interested in a much broader array of things, people, and events my newly found ability to walk now makes available to me.

"Graves studied values," says McIntosh.[151] "...the Gravesian mod-

---

150 www.claregraves.com

151 McIntosh is an earnest student of Clare Graves. He observes that "Graves's research revealed the existence of...familiar stages of psychological development that had been discovered by previous researchers. Graves's work, however, went farther than the other developmental psychologists in the way that it demonstrated all these sequentially emerging stages are themselves organized within a larger dynamic system. In other words not only did Graves's research reveal distinct, cross-cultural

A Universe Full of Magical Things

el of 'bio-psycho-social' evolution reveals how each stage is *a unified system that encompasses the development of values, worldviews, self-sense, belief systems, neurological activation, and a person's overall 'center of psychic gravity'* [emphasis added]... The Gravesian model shows how the individual and the culture at large (the micro and the macro) evolve together using the same stages and structures...[and] the Gravesian model exhibits a conceptual and geometric elegance that clearly reflects evolution's dialectical method of 'transcendence and inclusion'" (McIntosh, 2007, pp.32-33).

Each of the eight or so stages or levels identified by Spiral Dynamics is aligned with a primary set of values that shape how we think, how we see others, how we problem solve, what our motives are, and other aspects of our process at each level. Each level is more of a description of how we function rather than a description of what "type" we might be. Further we can function at one level in certain areas or circumstances of our lives, and at other levels in different circumstances. Certainly, when we are triggered emotionally, for example, we are less likely to be functioning in those moments from more sophisticated levels. In addition to solely individual dynamics and processes, Spiral Dynamics also describes the interactions between the individual and the culture, and the ways these eight stages play out in culture.[152]

In the last part of my earlier book, *Co-Creating a Brilliant Relationship: A Journey of Deepening Connection, Meaning, and Joy,*[153] I describe how we move developmentally through childhood and adolescence, to young, middle, and late adulthood. We have developmental tasks to master at each stage in a range of lines of development, one of

---

states of consciousness, it also showed how the same stages are related to each other in a *dialectic spiral of development*—a living system of evolution. Through his understanding of this larger system of development, Graves was able to discern how each discrete stage of development is shaped and formed by its relationship to the other stages... By recognizing not only the systemic nature of each individual stage, but also how all the stages were themselves related in a larger encompassing system, Graves was able to convincingly demonstrate how these levels of development in consciousness are actually a recapitulation of the stages of human history" (McIntosh, 2007, p. 32).

152 I describe Spiral Dynamics as well as Susanne Cook-Greuter's self-identity model in more detail on my website, www.DavidAYeats.com.

153 Yeats, David A., *Co-Creating a Brilliant Relationship: A Journey of Deepening Connection, Meaning, and Joy,* Outskirts Press, Parker, CO, 2014. See especially pp. 267-300. I extrapolated from the Spiral Dynamics model, along with Suzanne Cook-Greuter's research on self-identity, and worked out a simplified schema of sequential development of values in human growth, and tried to demonstrate the maturity of consciousness each level represents.

which is the development of our values, which as McIntosh, Graves, Beck and Cowan, and Wilber would argue, is a key line of development for individual *and* cultural evolution of consciousness.

The stage of our current life experience and our current life conditions determine our needs, and so, significantly determine our values.[154] Very simply stated, humans begin with a deep dependency on caregivers, (Waking Self), then move to a second stage of exploring our autonomy, (Budding Self). We do this in a variety of ways into adolescence, where we gradually adopt a mentality of self-concern premised on the idea of competition: everyone is competing, so I should take what I can get before others do, and as everyone else is doing, (Calculating Self). Sooner or later I realize I'll be more successful at getting what I need if I align with similar-minded others, and I develop a strong group mentality and identity, and add cooperative action to my skill-set, (Role-Playing Self). At some point I may begin to feel I have greater expertise or different needs than my group, so I may explore striking out again on my own, much as an employee may leave a job and move toward an entrepreneur lifestyle, (Prospering Self). Some begin to move away from these instrumental values orientations, seeing that everyone has a right to have opportunity and to strive toward happiness, (intrinsic human values) and we may begin to align with more pluralistic and inclusive values in our dealings in the world and with others, (Perceptive Self). From here a very few may move on to more spiritually-driven values perspectives, (Authentic, Connected, and Expansive Selves).[155]

You can see from this brief outline, perhaps, how and why the

---

154 This exploration incorporates data and perspectives from research with regard to a variety of lines of development particularly referencing Spiral Dynamics and Susanne Cook-Greuter's Developmental Models, along with insights from Ken Wilber's *A Theory of Everything*, (Shambhala, Boulder CO, 2000). Implicit in the stage descriptions are information from cognitive (Piaget), moral (Kohlberg and Gilligan), ethical (Fowler), and psycho-social (Erickson), among other stages/lines of development, and Wilber's integral perspective. Again, see www.DavidAYeats.com for summaries of the Spiral Dynamics and Susanne Cook-Greuter developmental models. See also Beck, Don Edward, and Cowan, Chris C., *Spiral Dynamics: Mastering Values, Leadership, and Change,* Blackwell, Malden MA, 1996; and Beck, Don Edward, *Spiral Dynamics Integral: Learn to Master the Memetic Codes of Human Behavior,* Sounds True, Boulder CO, 2006. Finally, see resources at www.cook-greuter.com.

155 In contrast to earlier developmental dalliances with religions and other "one right way" moralities based in "survival," these later developmental levels disidentify with the "story" we are living—the game we are playing—and are thus oriented toward a spirituality based in "being" that orients one's awareness to deeper realities, including appreciating the intelligent Cosmic Presence that pervades the universe.

values we hold in our lives evolve and develop in a sequential way, one stage or level flowing from a previous one. This occurs individually and culturally, and it occurs externally in the world, internally with our conscious relationship with our own being, and intersubjectively in our relationships with others and the world.

According to the Spiral Dynamics model, most folks in first world countries have developed as far as self-concerned (Calculating Self), group-identified (Role-Playing Selves)[156] or entrepreneurial (Prospering Selves). Culturally, especially in these days of instability and upheaval, it would be hard to imagine we have progressed beyond adolescence. So, as a culture or country, we are acting out of predominantly self-concerned (Calculating Self) values. The challenging realities we are reckoning with today suggest it will be difficult, as unquestionably necessary as it is, to move to higher stages of conscious awareness. On the positive side, it is out of life conditions such as these, that old paradigms seem more and more unworkable, which enhances the possibility that new, more mature values paradigms may result.[157]

In a variety of arenas in our lives, we can see the dynamic processes of evolution and emergence transforming our world. The cutting edge of that development is defined by our continually evolving self-reflective consciousness, our native human consciousness. At the most inclusive and holistic reaches of this native human consciousness, lies a consideration of the idea that consciousness is individual, variable, and pervasive, as well as collective and integral. There is only, according to

---

156 My simplified values schema synthesizes a number of developmental models. Here is the complete schema, (from Chapter 10, *Co-Creating a Brilliant Relationship: A Journey of Deepening Connection, Meaning, and Joy):*

Stages of Self and Modes of Action

Preparatory Movement: Waking Self – Arousers Budding Self – Learners Calculating Self – Contenders
Executive Movement: Role Playing Self – Joiners Prospering Self – Promoters Perceptive Self – Sharers
Emergent Movement: Authentic Self – Authors Connected Self – Allowers Expansive Self – Abounders

157 McIntosh amplifies on the significance of Graves' work in articulating the structures and dynamics of human consciousness as reflected in values emergence: "Graves is… most significant…because of his clear recognition of the systemic nature of the spiral of development…What makes Graves so important is not the data he accumulated, but his *interpretation* of the data. Graves achieved an unprecedented understanding of the interdependence of life conditions and the distinct values-based worldviews that arise in response to these conditions. His insights into the co-creative relationship between the stages, the 'bio-psycho-social' character of human development, and the recurring, spiraling nature of the dialectic [interactive and conflicting: tense] evolution of consciousness also represent significant contributions to our understanding of the internal universe" (McIntosh, 2007, p. 186).

---

quantum physics, "one interconnected being," and if we take that to be true, there is only, finally, one consciousness. It may be that our individual conscious awareness is like a single tiny mirror piece on a mirror ball filled with such pieces. We see what others see, and we don't, which leaves each of us with our own subjective version of reality. In the end, our collective consciousness is much more than our individual subjective states of awareness, and, in the end, we're in the game together.

Inspired to a great degree by the work of Claire Graves, the cutting edge contemporary philosopher, Ken Wilber, in his book, *A Theory of Everything*, crisply summarized the ways that human consciousness unfolds: "…Development, for the most part involves *decreasing* narcissism and *increasing* consciousness, or the ability to take other people, places, and things into account and thus increasingly extend care to each…

"Three general stages are quite common for most forms of development. They are known by many names, such as preconventional, conventional, and postconventional; or egocentric, sociocentric, and worldcentric, or 'me,' 'us,' and 'all of us'" (Wilber, 2000, pp. 18-19).

Wilber continues, "These three general stages—egocentric to ethnocentric to worldcentric—are of course just a simple summary of the many unfolding waves of consciousness… Each developmental wave is a *decrease* in narcissism and an *increase* in consciousness (or an increase in the capacity to take deeper and wider perspectives into account).

"… In Spiral Dynamics, the preconventional stages are beige (archaic- instinctual), purple (magical-animistic), and red (egocentric). Although red is called egocentric, the first two stages are even more egocentric (there is a steady decline in narcissism at each and every stage); it is just that red marks the culmination of the highly egocentric and preconventional realms and is now able to act this out forcefully. At the next stage (blue, conformist rule), the narcissism is dispersed into the *group*—for example, it is not me but my country, that can do no wrong! This conventional/conformist stance lasts into orange (egoic-rational) which marks the beginning of the postconventional stages (green, yellow, and turquoise). These postconventional stages (especially orange and green) are marked by an intense scrutiny of the myths, conformist values, and ethnocentric biases that almost always inhabit the preconventional and conventional stages…

"Each stage of development," continues Wilber, "brings not only new capacities but the possibility of new disasters; not just novel potentials but novel pathologies; new strengths, new diseases. In evolution at large new emergent systems always faced new problems… Annoyingly there is a price to be paid for each increase in consciousness, and this dialectic of progress (good news, bad news) needs always be remembered. Still the point for now is that each unfolding wave of consciousness brings at least the possibility for a greater expanse of care, compassion, justice, and mercy on the way to an integral embrace" (Wilber, 2000, pp. 20-22).

It is worth noting here that a powerful implication of native human consciousness is that our interior world, our internal consciousness is and will continue to evolve, *a critical feature of the direction of future evolution and even emergence*. McIntosh remarks: "The accumulating facts of cosmological, biological, and cultural evolution are now giving rise to a deeper philosophical understanding of evolution that recognizes interiors—agency, sentience, subjectivity, consciousness, and mind—play a central role in the universe's unfolding development. This new integral philosophy is not fully recognized by the academic mainstream, but it is nevertheless grounded in science and its arguments are intellectually rigorous…Consciousness can evolve in a wide variety of ways. It can be raised or evolved by increasing empathy and compassion, by cultivating knowledge, understanding and forgiveness, and by building political will and the determination to achieve social and environmental justice. Consciousness can also be raised by enlarging people's estimates of their own self-interest, by expanding their notions of what constitutes 'the good life,' and by persuading them to appreciate new forms of beauty and truth" (McIntosh, 2012, p. xxii).

Our capacity for self-reflective consciousness means we can set goals, act intentionally and deliberately in line with our values, and we can co-author with our universe how our lives, relationships, and our culture grows and changes. The resulting force of our co-emerging vision as a species has the power to create the outcome, energy, attitudes, relationships, and ecosystem we are envisioning. *What we focus on is what we get.* It is not just the universe doing its thing: it's *us* and the universe shaping our future and our world in participatory partnership. This is the most significant effect of the emergence of humanity and our capacity for self-reflective consciousness.

INTERSUBJECTIVE CONSCIOUSNESS

Specifically, "the quality of our interconnectedness" is *the primary vehicle* by which we can partner with the universe to move our collective native human consciousness forward.

"When we look at human culture and ask, 'what is it that is actually evolving?' we can see that it's the quality and quantity of connections between people, taking the form of shared meanings and experiences, agreements, relationships, and groups of relationships—these constitute what we might call the organisms of cultural evolution," says McIntosh (McIntosh, 2011, p. 18).

"This cultural domain of evolution," he continues, "...is known as the *intersubjective*—it exists between subjects" (McIntosh, 2007, p. 18). It is a subject-to-subject transaction, between two or more beings.

Consciousness *itself,* says de Quincey "is more than something objective or subjective—it is *intersubjective*" (de Quincey, 2005, p. 173). And that means that consciousness exists within conscious subjects, and, too, *outside* of the bodies of the engaging subjects—"in between us."

McIntosh points out that "...these agreements and relationships that are the structures of cultural evolution have their existence in the internal universe. That is, the substance of a human relationship is the experience that is in our minds and also in the minds of others. Relationships exist in the internal space 'in between us,' not wholly in our minds and not wholly in the minds of those with whom we are related, but mutually inside both of our minds, and often simultaneously. These relationship structures are partially independent from our individual subjective consciousness, but at the same time internal and invisible.

"These systems of human relationships are not 'in the air,' nor are they merely in our minds," he clarifies. "They exist in the intersubjective domain of the internal universe" (McIntosh, 2007, pp. 18-19).

Thus, in keeping with the lessons of quantum theory, we find again evidence that the universe is a conscious entity, best understood as dynamic movement, action, and events, as a process, as mind. Intersubjective consciousness is not entirely embodied, and yet it is fundamental to understanding who we are in space-time. It is the language of humanity, and it is the story of humanity. Yet aspects of intersubjective consciousness "exist" nowhere in what we experience as the physical realm, and have no physical correlates. Hoffman clarifies as well that "intersubjective agreement does not mean objective truth. It just means intersubjective agreement. And that's it. And yet, we have the psychological proclivity to say that since we all agree that it must be true. Not at all" (Hoffman, 2017, *Simulation*)[158]

De Quincey explains insightfully: "By looking at the problems from the very different perspective of *process* (where minds and bodies are understood as *temporal* relations rather than merely a spatial configurations) …a whole new array of possibilities open up for mind-body, mind-mind, and self-agency solutions…

"But this shift from substance to process is so counter-intuitive ('counter-habitual' is more accurate), so much against the grain of our usual ways of thinking and visualizing that it often takes people some time for people to 'get' it. Part of the problem is that our language is so steeped in substance-thinking—as if the world were a collection of *things* that move about in different configurations and relationships and not a series or network of *events* or processes out of which apparently individual 'things' or 'objects' arise and vanish…

"When we switch to thinking of minds and bodies in terms of process—as patterns of events in time—the relationship between matter and mind becomes much easier to comprehend… the way opens for an equally radical epistemological shift—an opening to solve the problem of other minds through the new understanding of subjectivity and intersubjectivity" (de Quincey, 2002, p. 100).

Adding to this, McIntosh says, "The relationships found in the intersubjective domain encompass more than our personal relationships with family, friends, and colleagues, they also include what might call

---

158 Hoffman continues, "That's another elementary logical mistake. The way to think about physical objects are not an indication of truth. They're just a convenient fiction, a standardized way of communicating to us what we need to stay alive" (Hoffman, 2017, *Simulation*).

"indirect" relationships—relationships with our favorite authors, artists, musicians, and public figures, living and dead. Indirect relationships can be remote in space and time—they do not require direct contact for real-time communication—yet such relationships can be very significant to our consciousness. Meaningful relationships need not be directly personal to move us; we can engage in meaningful relationships with our heroes by simply allowing their words, deeds, or art to communicate with us in the present. As long as there's communication (even one-way communication), there is a relationship" (McIntosh, 2007, p. 20).

"The most common approach to the study of consciousness in philosophy and science (at least in Western academic disciplines) is from the *third-person* perspective;" explains de Quincey, " that is, consciousness is studied as an object like other objects in the world—such as stones or stars, fishes or foxes, bones and brains. But the scientific study of consciousness as an object completely misses its essential nature...

"Some Western philosophers, and mystics throughout the world, recognize that the most significant characteristic of consciousness is its subjectivity; it should be studied from the *first-person* perspective."

"The *second-person* perspective is a third alternative. In this view, consciousness is neither an 'it' nor merely an 'I.' It involves *you*, or some other second person—it involves dialogic relations between two or more experiencing subjects. From this perspective, consciousness arises (or at least is informed and altered) when two or more subjects encounter each other and participate in some way with each other's being" (de Quincey, 2002, p. 176) (as McIntosh described in his preceding comment).

Offering a chronological perspective, de Quincey notes: "Historically the notion of the 'individual' as an autonomous self—one that could separate from the collective or community—is itself an evolutionary phenomenon. Prior to the time of Alexander the Great, from Homer down to Aristotle, the 'individual' was identified with the group or city state. At this stage, consciousness was pre-personal...a pre-individualized collective 'self'... Consciousness, 'knowing with,' was group consciousness, where members of the group had no sense of individual self-identity. Their identity was with the tribe or group."

De Quincey continues, making an insightful point, "Just as it was almost impossible for the average citizen prior to Alexander's empire to experience individual self-identity (pre-personal)—they just didn't *notice* the personal quality of consciousness embedded in the group—it has been almost impossible for the average individual in contemporary society to experience intersubjectivity at the level of transpersonal consciousness. Until now (and perhaps even still), we have been too embedded in our personal consciousness, in our Cartesian-Enlightenment 'lone ranger' individualism, to *notice* that *the deeper reality or grounding of our consciousness is the intersubjective matrix of interdependent relationships,*" [emphasis added], a matrix that shapes and sculpts our very sense of self.

"I'm proposing that a crucial aspect of the widely anticipated 'new paradigm'" he continues, "—a worldview of nonlocal interdependence—may be the emergence of *transpersonal* awareness of our deep intersubjective nature... It is becoming less and less easy to deny our deep interconnectedness" (de Quincey, 2002, pp. 213-214).

After distinguishing among first-, second-, and third-person points of view, de Quincey then goes on to point toward the ultimate goal of the evolutionary unfolding of the universe: "Standard third-person inquiry leads to a science of external bodies, first-person inquiry to an interior science of the mind, while second-person engagement leads to *a communal science of the heart.* Whereas the ultimate ideal of objective knowledge is control, and the ultimate ideal of subjective knowledge is peace, the ultimate ideal of intersubjective knowledge is relationship— and, dare I say it, love" (de Quincey, p. 180).

And why, exactly, is love the ultimate objective of the evolutionary unfolding of the universe? McIntosh concludes: "As evolution advances and organisms become more complex, this increasing complexity results in an increasing interdependence among the system's components. In other words, the more complex a system becomes, the more it needs to hold together its complex parts in ways that require a significant degree of integration. For a system or structure to maintain integrity in the face of greater complexity, it must become more unified..." (McIntosh, 2007, p. 277). [159] As we shift our perceptions from relating

---

159 What precisely becomes more unified as complexity grows? Rupert Sheldrake describes any "self-

to separate things to relating to unified aspects of our own being, treating ourselves with regard and reverence comes to mean loving all of it as aspects of our own precious being.

Finally, de Quincey, after describing what is truly an evolutionary unfolding from third-person to first-person to second-person, arrives at this perception: "Ultimately, when the experience of self expands to encompass the entire realm of being, as the mystics tell us, the distinction between individual self and cosmic Self, or the Cosmic I, disappears. The entire cosmos, then, resonates with meaning" (de Quincey, 2005, p. 1020. In the end, subjectivity and intersubjectivity are, after all, the only things that are not virtual, not merely representational, in this "physical" reality.[160]

---

organizing systems including molecules, cells, tissues, organs, organisms, societies and minds [as] made up of nested hierarchies or holarchies [called] holons" (Sheldrake, 2012, pp. 99-100). McIntosh teases this idea out: "The idea of a holon was first proposed by Hungarian thinker Arthur Koestler as a way to describe the basic units of organization in biological and social systems. Koestler observed that in both living organisms and social organizations, entirely self-supporting, noninteracting entities did not exist. In every domain of evolution there could be found neither simple holons nor simple parts because every entity subject to evolution was simultaneously a whole and a part. Koestler recognize how these whole-parts—systems embedded in systems—were organized in natural hierarchies or "holarchies" (hierarchies of holons), with every holarchical line of evolutionary development itself behaving as a self-organizing dynamic system. Koestler also saw how these naturally occurring holarchic systems emerged through the process of transcendence and inclusion—molecules transcend and include atoms and cells transcend and include molecules, organisms transcend and include cells. This concept of holons and holarchy arose as a way of modeling these complex systems…." (McIntosh, 2007, p. 219).

160 Eben Alexander: "Many might wonder just *why* we have shared awareness of a common world, since this shared reality seems to strongly support the independent existence of the physical world. This shared 'reality' is, in fact, the basis of our earthly experience. The consensus reality of the physical realm serves as a convenient stage for the unfolding of the events of our lives, even though, like the old Western town facades that Hollywood used for filming, *it does not exist as such*. In many ways, our jointly perceived physical world serves the same role as our bodies play in providing a focus of our here-and-now experience…

"…Of course, we agree that our perceptions match similar perceptions we've had in the past because we agree on definitions and understandings as they evolved. Thus our language is coded to mean certain things in certain situations, but we can only agree that those qualia [our subjective and qualitative aspects of our perception] remain stable to us in our ongoing perception of the world. We continue to face the problem of not knowing or being able to compare the contents and qualia of our perceptions with those of others. We are left merely to assume that we have some agreement in our perceptions.

"Many view the objective scientific method as the only pathway to truth, and believe that subjective individual experience counts for nothing in the current era of evidence-based conclusions. In fact all of human knowing is deeply rooted in personal experience (including that of the scientist performing the experiment, so that the *subjective* experience of the scientist, indeed all of us, must be viewed as integral…

"…a tremendous amount of information conveyed through subjective experience surpasses the knowledge available through 'objective' assessment alone—in fact, it is challenging to assert that unbiased objectivity can even exist.

A Universe Full of Magical Things

John Stewart observes the complexifying of the developmental movement toward self-reflective and intersubjective dynamics in this way: "Evolution progresses towards greater cooperation by discovering ways to build cooperative organizations out of [individual, personal] components that are self-interested. It has done so repeatedly throughout the history of life on earth" (Stewart, 2000, p. 10).

McIntosh discerns, lastly, that "complexity is in the direction of transcendence, and unity is in the direction of inclusion" (McIntosh, 2007, p.282). Intersubjective consciousness, it appears, is the cutting-edge dynamic underlying both.

"In fact, we never have access to any 'objective' reality. There seems to be a consensus reality that we assume to be the external world, including all beings in addition to ourselves, but we must be cautious not to assume too much about an objective reality outside of what is subjectively verifiable to us" (Alexander, 2017, p. 56-58).

Chapter Eleven:

THE SOUL OF OUR UNIVERSE— A COSMIC PRESENCE

BRINGING IT ALL TOGETHER

"…Everyone who is seriously involved in the pursuit of science becomes convinced that some spirit is manifest in the laws of the universe, one that is vastly superior to that of man. In this way the pursuit of science leads to a religious feeling of a special sort, which is surely quite different from the religiosity of someone more naive."

—Albert Einstein (Calaprice, 2002, p. 129).

"Constantly think of the universe as one living creature, embracing one being and one soul; how all is absorbed into the one consciousness of this living creature; how it compasses all things with a single purpose, and how all things work together to cause all that comes to pass, and their wonderful web and texture."

—Marcus Aurelius (Aurelius, 2015).

*"Consciousness is the light which illumines
those things on which it shines."*

—K. Ramakrishna Rao (Rao, 2011, p. 352).

"I am the one millions of suns shine on."

—Julian Yeats, ("Timeless Being," *Higher Vibe*, 2018).

We have seen that humanity's examination and exploration of the universe has led us down paths we could never have anticipated or predicted. Our physicality is not a chemical reaction as we have thought, but rather we are "an energetic charge," as Lynne McTaggart has said. "Human beings and all living things are a coalescence of energy in a field of energy connected to every other thing in the world. This energy field is the central engine of our being and our consciousness."

Thus we move from the traditional Materialistic Paradigm, to the Consciousness Paradigm that enrichens and deepens our insights about our cosmos and our existence.

We have also imagined that we existed in a sprawling universe that was constant, never really changing, that could and would exist forever. But we have learned we are far from infinite, having had our beginning in what we call the big bang, about 14 billion years ago. Our universe is expanding outward, at increasing speeds, and at some time in the future, it appears we will burn ourselves out. Humanity is not a permanent feature; rather, along with everything else we can imagine on this physical plane, we are merely a transient story.

Yet, as our universe expands, it increases in complexity, it grows and it evolves, and we discover more and more of its richness. We see that it is diverse, it exhibits a capacity to know, and it can take deliberate action. It is conscious, it is alive, intelligent, it changes and matures, and it is agentic and emergent. We have learned, in contradiction to everything our senses have indicated, that our world is made up of processes and events, magical things, but not material things.

Our science has been the arena where we uncover who and what we are, as well as all the various features of our seemingly hard and physical world. We have made great strides, and about 400 years ago, we concluded that there seems to be a predictability in the world, that things were determined in cause-and-effect ways. The world resembled a mechanical creation, a kind of stopwatch with its mechanisms and gears, and which when powered by the wizened magic of science, moved in unison and revealed time to us. The world seemed to be a machine. We came to believe—believe, not *know*—that "all events that ever happened anywhere at any time were wholly determined by prior mechanical causes," as Christian de Quincey describes. "All causality, therefore, was to be explained in terms of an unbroken chain of mechanical causes and effects. The present was wholly determined by the past, and the future likewise would be merely an extension of the same deterministic chain of events. There was no room for a nonmaterial, nonmechanical mind to step in and influence the causal chain."

As we have seen, and as Rupert Sheldrake described it, "This belief system became dominant within science in the late nineteenth century, and is now taken for granted. Many scientists are unaware that materialism is an assumption: they simply think of it as science, or the scientific view of reality, or the scientific worldview. They are not actually taught about it, or given a chance to discuss it. They absorb it by a kind of intellectual osmosis." Our sensory experience matches up with this cause and effect, deterministic, and mechanical view of how our world operates, which is why we are seduced into seeing this view dogmatically rather than as an undocumented belief. Seeing everything as physical, hard, and material, it is easy to conclude that everything *is* matter, and that everything in our experiences is a result of matter. It is, therefore, easy to conclude that existence *is* matter.

But, as I noted earlier, "the primacy of matter is treated as an unarguable and proven fact by mainstream science, *although no specific proof has ever been shown*. Mainstream science prefers to think matter emerged first in the universe because it is consistent with our experience of a physical world which we can examine."

Yet, ironically, it was this very discipline that, because of its persistent poking and prodding and its propensity to look under every material rock, we have been introduced to deeper, far more mysterious, incredulous, and bewildering conceptions of this experience we are having. In our quest to understand the most minuscule aspects of our world we discovered that the behavior of atomic and subatomic particles, this quantum theory, did not match the billiard ball reality of mechanistic physics. What occurred at the quantum level, in fact, turned out to be both not in line with our lived sensory experiences, but it also contradicted all of the scientific conclusions about our world that we had come to believe.

Yet, it worked. Every time. Without exception. As we've seen, its application in our technology has resulted in sophisticated possibilities far beyond what mechanistic science could offer. This new physics of quantum dynamics underlies our world at the deepest level. The macro level of the world we experience is built on the micro level of quantum dynamics—as surely as a massive skyscraper is built out of truckloads of bricks and mortar. And so, the macro world must be subject to the same physical laws as the micro world out of which it has been assembled. Indeed, it is not conceivable that anything might be constructed out of materials that when complete it no longer contains! So, in totality, the objects of our world are bound by the same laws and rules and dynamics as those of the micro world.

No need to shoot the messenger, but here's the conclusion: science acknowledges quantum theory as the underlying reality of our existence, and quantum theory says the physical world is virtual, not real in itself, because there is nothing that exists that is physical. What each of us perceive, uniquely and subjectively, is an illusion: what looks physical to us is not ultimately material or physical at all. So all we can say is, all we can call "real" are our perceptions. And perception itself is synonymous with our personal and subjective consciousness. So it seems we have to conclude that all that is "real" is our consciousness. Walker thus concludes that the data quantum physics arrives at "leads us to the incredible conclusion that *mind, or consciousness, affects matter*" (Walker, 2000, p. 95).

I believe we have no reasonable alternative but to accept that as true. Robert Lanza says it this way: "It's all only in the mind... the universe exists nowhere else" (Lanza, 2009, p. 167). [161]

---

161 Recent research in neurobiology ratifies the power of the brain to generate perceptions that don't match the reality. Pain studies show that one can have a real experience of pain even after one's body has healed from an injury, because of entrained neural pathways, and also that levels of pain can change simply as a result of suggestion. Search, for examples, "Lorimer Moseley" or "neural pain pathways" on You Tube. The experience of pain, says Moseley, occurs in the brain and nowhere else.

One can speculate that this is a solipsistic point of view, that knowledge outside of one's own mind might not exist *outside of* one's own mind.[162] Or one might contend that each of us subjectively sees the elephant, but from uniquely different perspectives. If there is consciousness in some form "all the way down," one might conclude that even a lowly quark perceives and participates.

Each of the contentions of materialistic science is premised on the assumption that somehow things are separate, and quantum theory demonstrates that everything is one thing: it's all interconnected. Thus, the universe, all that is, perceives itself.

And for THAT to be so, the universe must be conscious and alive. We have seen that observation, a capacity to know, is required for existence. Elisabet Sahtouris' perspective is that "We can construct a new scientific model...that takes into account the entire gamut of human experience and recognizes the cosmos as fundamentally conscious and alive … a holarchic, evolving, intelligent, process intrinsic to the cosmos itself."

We have now considered this consciousness and agency in the tiniest particle and in the flora and in the grand ecosystem that surrounds us, in the humble slime mold and in the ants and bees and birds and chimps and cows and coyotes and elephants and whales that keep us company. We have come to see that consciousness must be pervasive in the universe: "Everything that we talk about," declared Planck, "everything that we regard as existing, postulates consciousness" (Sullivan, January 25, 1931, "Interviews with").

If it's conscious, FAPP, it's alive. It's helpful to look at life and consciousness as on a continuum of greater or lesser complexity. In our conventions, we view life as coming on the scene at some (still) unspecified point. But that's deceptive. Even in a material worldview, we must conclude that matter and mind co-evolved together. We can say life

---

162 In its purest form, we would say, in solipsism, that, *everything* that exists exists only in my mind, including what we perceive as other humans. Yet we relate intersubjectively: not all that we are conscious of exists completely in me or in you, but an intersubjective consciousness somehow in between us, disembodied. Most reject the purest form, and make room for other beings with human consciousness.

Or we might understand a solipsistic interpretation as meaning everything exists not in my mind but the One Mind. As we are all interconnected this latter understanding makes the most sense, and presumably this is implicit in the biocentrism model.

and consciousness have always been in the universe, evolving over time in complexity. But it's more likely that it is not life and consciousness that have evolved, but rather our perceptions evolve, our personal consciousness evolves, resulting in greater appreciation of the complexities of consciousness and life, along with a growing capacity to value and act with agency at more emergent levels.

Aliveness runs throughout the one interconnected being that is the universe. The universe *is* a vital force. Its essence is alive, conscious, one. It is intelligent and agentic. It chooses. And, as humans, we participate in that choosing, in that evolving and emerging unfolding and complexifying and becoming, bringing our own self-reflective consciousness along, and envisioning how we might co-create toward a greater and greater expression of love and compassion and care.

Walker holds that "in their essential nature, quantum fluctuations are the stuff of consciousness and will." And he goes on to say, "We discover that in the beginning, there was the Quantum Mind, a first cause, itself time-independent and nonlocal that created space-time and matter/energy" (Walker, 2000, p. 326).

Again, as I stated earlier, "The current scientific understanding of the beginnings of our universe (the big bang and inflationary theories) say that it is the *quantum fluctuations* that were present in the *infinitesimal vacuum* which spawned the big bang: it is a *quantum presence* which germinated physical manifestation."

If we ponder this mindboggling conception of "quantum fluctuations" existing in the "infinitesimal vacuum" that preceded the big bang, we have to conclude that science has outpaced the data it actually possesses. Referring to an existence and a time before the big bang is only fanciful, because one, we know nothing about it, and two, we are using spatio-temporal (and symbolic, culturally-bound, and virtual) language that by definition employs what follows the big bang to describe what precedes it.

Being limited by language, as we are, the term "quantum fluctuations in the infinitesimal vacuum" is merely code for something that is unknown to science. We can speculate that quantum dynamics are at play in whatever is "earliest" in existence, and we can affirm its nonphysicality. Walker's using the term "quantum mind" is a nod to that speculation. I think we've persuasively argued for the primacy of consciousness, which is, or at least heralds, quantum fluctuation and quantum mind.

Acorns and apple seeds appear simple, and there is nothing observable or evident to suggest that each might complexify, or what each might complexify into. Yet, what results is inexplicable. So too, out of apparently simple beginnings, with nothing observable or discernable to suggest complexification, we have witnessed existential experience, intelligence, consciousness, mind, self-reflective awareness, life, and love inexplicably emerging. It can only be naïve and simplistic, drinking the Kool-Aid, to conclude that any of these results might be random and arbitrary, merely an outcome of chance and mechanical selection. Yet this is the miracle that reductive science, the great denier of miracles, would have us believe.

De Quincey points out that not all cultures assume such a reductive stance. He thinks that "Belief in the separation of mind and body may be at the root of many of our current problems in science, in medicine, and in our personal experiences of alienation from nature. This dualism is an intellectual inheritance peculiar to the West. It is not part of the metaphysics, cosmology, religion, philosophy, or mythology of many other traditions. The Chinese, for instance, do not make such a distinction—they have no word for 'mind' in their language. In their view, all things are formed by the interplay of *ch'i* (matter-energy) and *li* (organizing principle)" (de Quincey, 2005, p. 157).

The something-out-of-nothing "answer" to our presence here that conventional science feebly (and miraculously) offers is an enormously unsatisfying and disingenuous non-answer. Which leaves us with the far more credible miracle of a first cause, (or quantum mind or consciousness or God), which, in the language of virtual technology, might be called the "booter-uper." In the absence of any other contenders, it seems like some sort of acknowledgement or homage to a first cause makes sense rationally.

Now here we move to the question of meaning or purpose. In acknowledging the virtual reality that it truly seems is what we are experiencing, we've seen that all that we *know* to be "real" is our very consciousness itself. It is a consciousness we share with all other conscious entities—human, at least, or perhaps "non-human persons" as well, as Harvey proposes, or even rocks, as Bohm believes, or particles, as Dyson suggests. Like mirror panels on a disco ball, each of us reflects some bit of the totality of consciousness, intersubjectively similar and subjectively unique.

But what is real is not the reflection we see in those mirror panels. Likewise, what is real is not a reflection of the "physicality" of our brain. Recall Donald Hoffman's elucidation: "I'm emphasizing the larger lesson of quantum mechanics: Neurons, brains, space ... these are just symbols we use, they're not real. It's not that there's a classical brain that does some quantum magic. It's that there's no brain! Quantum mechanics says that classical objects — including brains — don't exist.

"So this is a far more radical claim about the nature of reality and does not involve the brain pulling off some tricky quantum computation" (Gefter, "The case against," April 4, 2016).

As I commented earlier, "Quantum realism, virtual reality, contends: what we experience through our senses are symbols of an underlying reality, perceived through our conscious observation, but not the underlying reality itself. Within a virtual game, one's experience is real, but applies only within the game: it is not the total and complete reality, which would include the player, the programmer/booter-uper, and any other attributes of those broader realities."

EVOLUTION'S PURPOSE: EVOLUTIONARY SPIRITUALITY

In the Consciousness Paradigm we grow and change. Growth and change appear to be pretty central feature of our story, both individually and collectively. As we've seen with developmental research in general, and with Spiral Dynamics particularly, the path of our growth is deeply intertwined with our life conditions and our values. As we evolve, our values evolve in a sequential manner, and the further along the continuum we go, the greater will be our focus on care and compassion, and the more we experience inclusivity: our sense of "us" grows larger and the purview of our embrace expands.

Why is this? In fact, why is it that any of the aspects of evolution we've explored occur in what we might call a "lawful" way? Beyond that, why are there apparently predictable physical laws in the universe? It seems we can say there must be an underlying structure or maybe a grid of some kind, like the grid of spacetime, which is the dance floor for the dance. It all happens on the dance floor.

Or, it all happens within the confines and programming established for our virtual game, all orchestrated for us to experience our story. But here's a curious thought: perhaps there is reciprocal and ongoing communication between each of us mirror panels of consciousness and the Consciousness of our booter-uper.

David Bohm thought the explicate order of our lives was, in effect, the dance floor for our dance. But he also thought there was an implicate order, a level of information or consciousness we might at times gain access to, but for the most part was available to us in only delimited ways. Another version of the earthly and heavenly realms. But what is the nature of the connection between the two, how is it accomplished, and what's the purpose?

Perhaps the "brain," the earthly, physical "location" of our mirror panels of consciousness does not hold the content of our consciousness, but is the receiver, the information resource which connects us beyond the virtual game with the booter-uper? And what if the brain-as-receiver connects the booter-uper with us, like a two-way radio?

That would mean we have always-available two-way communication with the booter-uper, with Consciousness beyond the virtual game. So we could appreciate the brain-as-receiver, or the mirror-as-reflector, for providing a "physical" vehicle, like a radio or mirror, for that communication.

There is, then, a virtual reality game "grid," the dance floor, which continually plays out its program, and wherein we largely unconsciously play out our role and our script. But within the noise of that, we are able, with deliberate intention, to access the knowing and wisdom of the booter-uper. When we sit in nature, or reflect, or meditate. Or wonder. When we "listen to our inner wisdom." And this is a primary way for us to evolve, for us to develop toward higher values of greater inclusion, where, more and more, we can feel aligned and in tune with reality beyond our virtual game, and rather than fuss over our separateness, we can bask in our connectedness.

That is likely what the end game of our virtual story is about. Or, at least, it's congruent with all that has been written here. It's soothing and clarifying, it feels wise and good, and it offers a richness and abundance of purpose and meaning.

Yet the game seems to have a huge amount of both good and bad, which makes it hard and at times, oh-so-very-hard, for us living beings. It does seem, though, as we've learned, that a lot of that which we think of as bad is our personal subjective experience, and that, for another or others, what we experience as bad could be good. It does seem that maybe we need this contrast in our virtual game in order to grow toward the higher values we've described. Maybe that's why there's the game at all: to catalyze us toward a greater consciousness and a greater good that the booter-uper has planned.

John Stewart, in his book, *Evolution's Arrow,* offers a powerful perspective on evolution and humanity's role in its dynamic. He suggests that in our personal developmental growth, we might reach a stage wherein we develop an attention to the evolutionary dynamic which seems so central to understanding the virtual game we are playing. He notes that "It is not easy to escape the control exercised over us by our existing motivations, beliefs and values…if we have little evolutionary awareness, the evolutionary needs of far-distant generations will not count for much against our more immediate needs for food, sex, money, power, and social status.

"But evolutionary awareness can change this. It tends to produce individuals who place more value on the evolutionary success of future generations, and less on the gratification of their own immediate urges and needs. Individuals can see that it is the on-going and evolving population of organisms that are important to evolution, not any particular individual. Individuals are significant only to the extent that they contribute to the success of the ongoing evolutionary process."

In the big picture, he declares, "Evolution ensures that individuals that are not self-evolving are temporary. Individuals are evolutionary experiments…

"Evolutionary awareness shows us that it is an illusion to see ourselves or other individuals as distinct and separate entities. Individuals are inextricably part of an ongoing evolutionary process. They can have no existence without that process. They are born out of it, and can have on-going effects only through what they return to it. In our mental processes, we can separate individuals out from the ongoing process and consider them as independent entities. But in reality, they are never separate. Without an ongoing population that reproduces and evolves through time, there can be no individuals... Individual organisms such as ourselves are always parts of larger evolving processes" (Stewart, 2000, pp. 315-316). Again, we are interconnected. We are one.

Stewart points out that "the advantages of cooperation can be expected to drive progressive evolution wherever life emerges. On any planet where life evolves, evolution can be expected to produce cooperation over wider and wider scales, as it is on earth," (Stewart, 2000 p. 11), and he highlights, "It is not only through increase in cooperation that evolution progresses. It also progresses through increases in the ability of living processes to adapt and evolve...

"Evolution itself evolves, and living processes get smarter at evolving" (Stewart, 2000 p. 12). He explains that the successes of the past required cooperation, as Lipton similarly described, and in so doing, indicates our path as participators in evolution—the path to our continued success: "We are cooperators that are made of cooperators that are made of cooperators, and so on. We are cooperation all the way down.

"...Cooperation is widespread only because it has been able to defeat non-cooperation in evolutionary struggles" (Stewart, 2000 p. 43).

"We can live our lives as if we are separate from the on-going evolutionary process, and make no conscious contribution to it," Stewart continues. "But individuals with evolutionary awareness will not find meaning or purpose in such a life. They will find no meaning in a life spent vigorously and energetically seeking the satisfaction of their pre-existing material and emotional urges. A life dedicated to the pursuit of money, sex, power, and social status within our current social structures is a wasted life from the perspective of evolutionary awareness. Meaning can come only from taking whatever opportunities we have to contribute to the on-going evolutionary processes that will outlive us" (Stewart, 2000, pp. 317-318).

"When life evolves successfully on the planet it will eventually produce a cooperative organisation on the scale of the planet. The planetary organisation will be established by external management that supports cooperation and suppresses destructive competition within the organisation. Members of planetary organisation will be aware of the direction of evolution, and will consciously design and structure their social organisation so that it pursues evolutionary objectives" (Stewart, 2000, p. 282).

Of course, we are aware of great challenges today in the pursuit of any evolutionary objectives; from toxic overpopulation, to climate degradation, to economic warfare in our and other democracies, to the ethical encounters presented by "the pursuit of money, sex, power, and social status" and epidemic greed—all of which we quickly, consciously, and collaboratively must find a way to address.

In my work with psychotherapy clients, I've seen that those who experience the deepest levels of joy, satisfaction, and meaning are those that are not motivated by their moment to moment feelings—the motivating force for most of us—but those who take time to access, identify, and act on their wisdom, irrespective of whether it matches what they are feeling in this moment. So I encourage clients, at the very least to wonder what their inner wisdom has to say to them. Evolutionary awareness minimally demands this degree of self-awareness.

Freeman Dyson's perspective maintains that as long as our universe endures, it will continue to evolve, with or without humanity's participation. The great process of the expansion of consciousness will persist: "The infiltration of mind into the universe will not be permanently halted by any catastrophe or by any barrier that I can imagine. If our species does not choose to lead the way, others will do so, or may have already done so. If our species is extinguished, others will be wiser or luckier. Mind is patient. Mind has waited for three billion years on this planet before composing its first string quartet. It may have to wait another three billion years before it spreads all over the galaxy. I do not expect that it will have to wait so long. But if necessary, it will wait. The universe is like a fertile soil spread out all around us, ready for the seeds of mind to sprout and grow. Ultimately, late or soon, mind will come into his heritage" (Dyson, 1988, p. 118).

McIntosh suggests our best hope resides in our proceeding from a viewpoint that makes room for a spiritual process as central to our growth and unfolding: As our worldview grows wider and deeper, it becomes increasingly evident that the story of evolution in nature, self, and culture has unmistakable spiritual implications, says McIntosh. "Indeed, it is hard to deny that evolution's advance from the isolated atoms of hydrogen gas to the blue jewel planet earth, then from single-celled prokaryotes to the first self-conscious humans, and then finally from the first humans to the twenty-first century's global civilization, is the astonishing odyssey of development, a magnificent production of form and diversity, and unfathomable outpouring of creativity" (McIntosh, 2007, p. 217). All in service of deepening complexity, consciousness, compassion, cooperation and collaboration, and co-creating, and altogether in one interconnected emergence.

A COSMIC PRESENCE

*"God is what mind becomes when it has passed
beyond the scale of our comprehension."*

—Freeman Dyson, *Infinite in All Directions,* (Dyson, 1988, p. 119).

*"All are made of love and light.
All are full of love and light.
So pass the joy (love and light).
Pass the joy!"*

—Julian Yeats ("Pass the Joy," *Higher Vibe*, 2018).

Now, we finally arrive at The Quantum Mind Paradigm, a Mind that is beyond space, time, matter, and choice: a mind that transcends and includes all that is.

It seems true for me that my brain in the material world is not the container for what I experience, but rather, it is the receiver. It is the receiver of information that does not necessarily or completely exist in me, but which I can access.

It is the earthly tool by which I might access the heavenly.

It exists in the explicit order for the primary purpose of accessing and utilizing the information of the implicate order.

It is the means by which I, in my subjective and very personal consciousness, am able to connect with the greater and vaster underlying Consciousness, the Soul of the Universe, the Cosmic Presence, Spirit, of which I am a part.

My brain, and yours, participate in this virtual game, known as physical reality. I and we have things to accomplish in this game, including understanding at deeper values-levels what those accomplishments consist of. We observe and participate, and we do so as agents of the first cause, the booter-uper of this game.

We have ongoing access to the mind of this booter-uper, because of our consciousness, and because that consciousness is an aspect of the Soul of the Universe that not only pervades it all, but *is* it all.

The Cosmic Presence is everywhere, and it is right here, where I am. You can see me at the center of the universe. And you can see yourself at the center of the universe as well, because you are it and I am it and we are it and all is it. This Cosmic Presence, the Soul of the Universe, is all that is, and it exists within the game and beyond the game. The game itself exists so we can notice; it is a vehicle for us to begin to grok this Cosmic Presence, to sense it is who we are, and to do its bidding.

That which the Cosmic Presence, Spirit, is bidding us to do is something we discover personally, subjectively, and uniquely, but the path to discovery gets clearer the more we move up the developmental values spiral—in greater complexity as we go—toward greater inclusivity, because, duh, we are all one.

If we choose to, we can embrace the truth we've gleaned from quantum theory—that we participate in and co-create as part of one interconnected being. We evolve and emerge and grow more complex, more conscious, and more compassionate. And we can do so, if we choose, more deliberately, more mindfully, and more collaboratively.

McIntosh believes that our spiritual journey involves a more conscious and more deliberate attention to the "fundamental tenet of evolutionary spirituality...that the beautiful, the true, and the good represent the most intrinsic form of value humans can experience or create." (McIntosh, 2015, p. 53). Noticing, expressing gratitude for, and generating these values within ourselves and within our world are practical ways we can deepen our spirituality, and contribute to the evolution of these values in our world, he says.

Otherwise stated, humanity's work is to nurture compassion in the service of wholeness and unity. Individually, as Esther Hicks, the channeler of the collection of spirits known as Abraham, would counsel, your work is to align with Spirit, the Soul of the Universe, and do what you prefer. The outcome for us, as we are able to align our preferences more and more with The Cosmic Presence is greater joy and greater well-being.

We live in a Universe Full of Magical Things, wherein nature is conscious and alive, and wherein all that exists on Earth is of nature: all is one. Let us, then, be mindful enough to tend to our well-being.

Epilogue

"The cosmos…is like a luminous symphony: a complete, flowing, and unfolding process of creation. Many players move together, bringing the different perspectives of different instruments, different expressions, different tempos and moods and movement; moments of silence contrasted with moments of great exuberance. Great passion and great common meaning congeal, infused with unique and diverse perception, sensation, experience, and elegance. Each aspect, each piece, separate and differentiated though they may be, are all in truth one facet, one melody, one harmonious stirring expression: one great essence."

—David A. Yeats (Yeats, 2014, p. 322).

"There are eight million stories in the naked city. This is one of them."

—from the 1948 TV Series, *The Naked City.*

Living from a place of self-knowledge and joy involves listening closely to your inner wisdom, mindfully choosing your wisdom over your feelings, and acting with integrity.

—David A. Yeats.

As you may have surmised, this has been a life-altering and consuming project for me. I'm still a bit perplexed about how I got into it, and so deep. But there it is. The net result is immensely grounding and satisfying for me, and I hope you'll take something of that with you for yourself.

The contrast between the material world and the quantum world is real, and it demands the kind of attention given here to try to make sense of it. I found it clarifying and helpful to speak of these things from the standpoint of paradigms, because a great deal of what gets debated is premised on the point of view one is coming from. The dominant Materialist Paradigm must be the starting place, and none can disagree that its premises are powerful, in that what we experience is a hard and physical material world, and it is a world we are compelled to deal with every day. It's no wonder that's science has struggled mightily to ratify a physical only version of reality. We know it succeeded very well until quantum theory came along, and undid the cherished "matter first," cause and effect, billiard ball mindset.

But quantum theory did come along, and with it the need to reckon with the information that, at the minutest levels, there is no such thing as matter at all. There is consciousness. Quantum said that all we can assert is real is our perceptions, each similar but not the same. Perception is based in interpretation, and we've learned that what we were interpreting as physical was an illusion. The Materialist Paradigm, as reassuring as it was, was not true.

Instead, we live an illusion. What we experience is a virtual reality, a game, a representation of what is real, but not the final reality; like playing on a holodeck. But the holodeck exists on a starship—thus there is something more than what we perceive in our holodramas. Our holodeck is a pretend place where things appear as hard and physical—but exist only within the holodeck universe. The holodeck contains the entirety of our physical virtual game, which we are taught to treat as real.

Now we can grok that *what we hold to be true, isn't*. There is no objectivity. Time and space and matter, even choice, are all illusory—not true as they present to us in our perceptions. What I experience is subjective, unique, and personal, and is not entirely the same as what you experience. The world of my experience is made up, like a programmer makes up a virtual game. All the events, avatars, and interactions have meaning, but only within the game. Concepts such as infinity have no meaning in a non-physical universe: every direction I look goes on until I no longer look.

What is real is only my individual observation, my personal perception, which is both the same and different than the perceptions of others, each an aspect of a greater consciousness. This greater consciousness is the programmer of our virtual game, the booter-uper. This programmer has also injected aspects of its greater consciousness into every facet of our game.

So our virtual game consists of everything we have seen and discovered and everything we will see and discover. Its venue is materiality; and we've agreed to live by the rules of matter. However, in the last five hundred or so years we have started to see that underlying materiality is awareness, and that fundamental consciousness might actually be the original reality that drives everything—because we don't live in mechanical billiard ball world, but rather in an ultimately non-physical world where we have a capacity to know, and a capacity to choose, in the context of the game.

Because we are not physical, but one interconnected consciousness, there is no separation between us. Separation too is an illusion. We are one in our virtual reality, and also, we are a part of the one consciousness with the programmer. The reasons we are here have been determined by the programmer, as well. Although in this world we each have a role to play, it is not prescribed: it is something we participate in creating with the programmer. It is our work to align with those reasons, as best we can understand what they are.

Traditional physics belongs to the Materialist Paradigm. It seeks to explore and understand and plumb the depths of physicality, and is true to that extent only. The Consciousness Paradigm, which we've just described, holds that there are no "its." All the "things" we perceive are virtual "packets of action," quanta, which make choices, and all we perceive is conscious, at least at a primordial level, because everything is constituted of choice-making "particles," the primal form reality takes on a physical level. Everything from buses to roads, and from skyscrapers to a grain of sand on the beach, are ultimately, from the Consciousness Paradigm perspective, the various forms that physical reality takes. Yet, beyond this materialist illusion, this materialist myth, is an aliveness that pervades everything. These "things" are consciousness manifested, representations pointing to a deeper and wholly nonphysical reality. In the Consciousness Paradigm, consciousness, life, does not originate from matter; matter is derivative from consciousness. Consciousness passes through and pervades all of matter.

Coming from our familiar physically experienced lives, the idea that we might live in a nonphysical and virtual reality must sound bizarre. Incongruously, and fatefully, as we've seen however, it is the proponents of "matter first," those who went looking for the tiniest of material things, who were the ones that discovered that at its tiniest, there is no such thing as matter at all. That is what science says. Matter exists only as swirling vortices of energic events, packet of action.

Ever since then, traditional scientists have been doing back flips trying to hold onto the belief that all things come from matter, while quantum theory's essential premise is that it's observation that creates reality, not matter.

So the Consciousness Paradigm replaces the Materialist Paradigm with the ascension of quantum mechanics. It is a story of each human's unique journey through a subjective, choice- making, evolving, growing, perceiving, and emerging toward more mature levels of consciousness toward an increasingly more inclusive and compassionate set of values for being in the world. If we are one, we ought to put the wellbeing of all of us first.

Every night, I try to step outside for five or ten minutes, and look at the sky. I am aware that who I am is a player in this material-based illusion game. Even as I recognize its seductive hallucination, I remember it's pretend, a dream, albeit alive throughout.

I look up at the heavens, past the clouds and the moon, the planets and the stars, as far as I can see. I think of our game as a bubble, which extends out as far as humanity might ever see. Here I am in this story—imaginary, though it doesn't seem that way—and engaging and joyous, painful and pleasurable, agitated and calm, angry and love-filled. I intend to look past it all, past the bubble and all it contains, and try to remember the Great Consciousness, the Quantum Mind, that is it all and transcends it all. I try to remember that what is real is beyond this virtual story, and carries such depths of meaning beyond all this. I commit to living from this place of awareness, and I commit to allowing it to be, just as it is, and to acceptance and non-judgment and compassion and gratitude and mindfulness. Then I go back inside to this life, this story that I am living.

But that is just my story. In the period of time humanity has been on the planet, it's estimated that there have been 110 billion to 113 billion other stories, all equally valid. 113 billion manifestations of consciousness. Like a mirror ball with 113 billion separate reflections of One Light. Part of my work is to respect and honor each of them. We all have a bit of the truth, and we all have our delusions. Whatever reason all these stories exist is not mine to judge. I assume that everyone is doing the best that they can, and that's it.

Just as each human story is time-bound and finite, so too is our cosmos. Our world is dying. Humanity is dying. There has never been a species on earth that has lasted forever, and even if we believed we could achieve it, ultimately, our universe will cool, and the story of human life with it.[163] So it's not how long we live, but how. Education is good, because we grow to be more conscious, as Spiral Dynamics has described. But as a species, we pay more attention to our interests and survival, so there will be limits to the power and influence of education.

---

163 In the Washington Post article, dated October 6, 2017, by Christopher Ingraham, entitled "We have a pretty good idea of when humans will go extinct," the author observed, "Mammalian species typically last around one million years before going extinct. You could argue that our species' intelligence gives us a survival edge over say, a mastodon or a rabbit, which could make us more likely to beat those odds...

"Neanderthal ancestors were around for only 300,000 years, while *Homo erectus* survived for about 1.6 million. They were smarter than the animals around them, but from a longevity standpoint they were completely unremarkable. Why should we be any different? Why should we be special?" (Ingraham, October 6, 2017, "We have").

Further, in the September 5th, 2018 article in the UK's *Express,* "Humans WILL go extinct and there's NO hope for survival: It's WHEN not if, biologists say," author Sean Martin (Martin, September 5, 2018,"Humans will"), reports it is not the first mass extinction throughout the planet's history—it is actually the sixth. According to biologists, just 0.01 percent of animals throughout history are living today. This means virtually all species will die off at one point or another. Without giving a timeframe on when it will happen, experts believe that humans' time will come one way or another, either through a natural disaster – such as the asteroid which wiped out the dinosaurs – or a man-made event like war.

"Luke Strotz, a post-doctoral researcher in invertebrate paleontology at the University of Kansas, in an article for *The Conversation* entitled "What makes some species more likely to go extinct?" wrote: "Death is the inescapable endpoint of life. And this is as true for species as it is for individuals. Estimates suggest 99.99 percent of all species that have ever lived are now extinct. All species that exist today – including human beings – will invariably go extinct at some point" (Strotz, April 22, 2018, "What makes").

We have to set that all aside, and live, each one of us, the best life we can, and follow whatever path is ours to follow, remembering that we are all stewards of Spirit.

I believe what is written here can contribute to how we better live, how we understand our existence here, and can help clarify what we do next. We are all in superposition, until we choose what's next. It's best it seems, if we trust the Universe. It is full of life and wisdom and intention, deliberateness, intelligence, choice, and creativity. And Love. It is, indeed, full of Magical Things!

References

Ackerman, Jennifer, 2016. *The Genius of Birds*. New York; Penguin Books.

Alexander, Eben, 2017. *Living in a Mindful Universe A Neurosurgeon's Journey into the Heart of Consciousness*. New York: Rodale.

Alexander, Eben, 2012. *Proof of Heaven: A Neurosurgeon's Journey into the Afterlife*. New York: Simon and Schuster.

Arnold, Jack, April 16, 1955. "Time is Just a Place." Science Fiction Theater TV show.

Aurelius, Marcus, 2015. *Meditations*. New York: Penguin Classics, Reissue edition.

Bacon, Francis, 1597. "On Aetheism" in *Essays: Counsels, Civil and Moral*. Public Domain.

Ball Philip, (2/17/2015). "Too Many Worlds: Nobody Knows What Happens Inside Quantum Experiments. So Why Are Some So Keen to Believe in Parallel Universes. *Aeon*, (Digital magazine) Retrieved from https://aeon.co/essays/is-the-many-worlds-hypothesis-just-a-fantasy.

Barrow, John D., Simon Conway Morris, Stephen J. Freeland, and Charles L. Harper, Jr., eds. 2008. *Fitness of the Cosmos for Life: Biochemistry and Fine-Tuning*. Cambridge: Cambridge University Press.

Barrow, John D., 2002. *The Constants of Nature: From Alpha to Omega—The Numbers that Encode the Deepest Secrets of the Universe.* New York: Pantheon Books.

Beck, Don Edward, 2006. *Spiral Dynamics Integral: Learning to Master the Memetic Codes of Human Behavior. Boulder CO:* Sounds True.

Beck, Don Edward, and Cowan, Chris C., 1996. *Spiral Dynamics: Mastering Values, Leadership, and Change.* Malden MA: Blackwell.

Bergson, Henri, 1998 (1907). *Creative Evolution: Humanity's Natural Creative Impulse.* Mineola, New York: Dover Publications.

Bichell, Rae Ellen, February 24, 2017. "Could a Bumblebee Learn to Play Fetch? Probably." *The Salt.* Retrieved from http://www.npr.org/sections/thesalt/2017/02/24/516532358/could-a-bumblebee-learn-to-play-fetch-probably.

Bohm, David, 1980. *Wholeness and the Implicate Order.* New York: Routledge and Kegan Paul.

Born Max, 1935. *The Restless Universe.* London, Great Britain: Blackie and Son, Limited.

Boyce, Neil Greenfield, August 30, 2016. "Their Masters' Voices: Dogs Understand Tone and Meaning of Words." *NPR.* Retrieved from https://www.npr.org/sections/health-shots/2016/08/30/491935800/their-masters-voices-dogs-understand-tone-and-meaning-of-words.

Brenkus, John, and Stern, Michael, (executive producers), 2009-2011. *The Known Universe.* National Geographic Channel series.

Bridle, Susan, Spring-Summer 2001. "Comprehensive Compassion: An Interview with Brian Swimme." *What Is Enlightenment?* No. 19, p.40.

Buffett, Jimmy and MacAnally, Mac ,1999. "Oysters and Pearls." *Beach House on The Moon*. [CD]. New York: Island Records.

Capra, Fritjof, 1982. *The Turning Point: Science, Society, and the Rising Culture*. New York: Bantam.

Carrington, Damian, October 29, 2018. "Humanity Has Wiped Out 60% of Animal Populations since 1970, Report Finds." *The Guardian*.

Casti, John, 1989. *Alternate Realities: Mathematical Models of Nature and Man*. New York: John Wiley & Sons.

Chomsky, Noam, 1972. *Language and Mind*. New York: Harcourt, Brace, Jovanovich.

Chopra, Deepak, Dec 06, 2017. "Artificial Intelligence Will Never Rival the Deep Complexity of the Human Mind." Retrieved from https://www.huffpost.com/entry/artificial-intelligence-human_b_10240122.

Chopra, Deepak, 2006. *Power, Freedom, and Grace: Living from the Source of Lasting Happiness*. San Rafael: Amber-Allen.

Chopra, Deepak, 2004. *The Book of Secrets: Unlocking the Hidden Dimensions of Your Life*. New York: Harmony Books.

Cole, K. C. (2001). *The Hole in the Universe*. New York: Harcourt Inc. As referenced in Whitworth, Brian, *Quantum Realism*, 2014.

Combs, Allen, 2002. *The Radiance of Being*. New York: Paragon House.Combs, Allan and Holland, Mark, 1996. *Synchronicity: Through the Eyes of Science, Myth, and The Trickster*. New York: The Marlowe Company.

Conway, John; Simon Kochen 2006. "The Free Will Theorem". *Foundations of Physics*. 36 (10): 1441. arXiv:quant-ph/0604079. Bibcode:2006FoPh...36.1441C. doi:10.1007/s10701-006-9068-6.

Cook-Greuter, Susanne. See "A Summary of Susanne Cook-Greuter's Developmental Model" summarized by David A. Yeats. Retrieved from http://www.davidayeats.com/A-Summary-of-Cook-Greuter-s-Developmental-Model.html.

Corning, Peter, 2005. *Holistic Darwinism: Synergy, Cybernetics, and the Bioeconomics of Evolution.* Chicago, University Press.

Crew, Bec, June 1, 2016. "The Case For Black Holes Being Nothing But Holograms Just Got Even Stronger." *Science Alert.* Retrieved from http://www.sciencealert.com/the-case-for-black-holes-being-nothing-but-holographic-images—just-got-stronger.

Crick, Francis, 1981. *Life Itself: Its Origin and Nature.* London: MacDonald and Co.

Crick, Francis,1966. *Of Molecules and Me.* Seattle: University of Washington Press,

Davies, Paul, 1998 *The Fifth Miracle: The Search for the Origin and Meaning of Life.* New York: Simon and Schuster.

Davies, Paul, 1982. *The Accidental Universe.* New York: Cambridge University Press.

Davies, Paul, 1984. *Superforce.* New York: Touchstone.

De Duve, Christian, 1984. *A Guided Tour of the Living Cell. 2 vols.* New York: Scientific American Library.

Denton, Michael, 2001 "Organisms and Machine: The Flawed Analogy," in Kurzweil, George F., and Ray. *Are We Spiritual Machines? Ray Kurzweil vs. the Critics of Strong A.I.*, Jay Richards, ed., Seattle: Discovery Institute.

Denver John, 1990. *The Flower That Shattered the Stone* (written by John Jarvis and Joe Henry).

Descartes, René, 1629 *The World.* Public domain.

De Quincey, Christian, 2015. *Synchronicity.* Retrieved from https:// video.search.yahoo.com/yhs/search?fr=yhs-itm-001&hsimp=yhs00 1&hspart=itm&p=christian+de+quincey+synchronicity#id=1&vid =845e64ebf81501e74b7ebb65cdc15160&action=click.

De Quincey, Christian, 2005. *Radical Knowing: Understanding Consciousness through Relationship.* Rochester VT: Park Street Press.

De Quincey, Christian, 2002. *Radical Nature: The Soul of Matter.* Rochester VT: Park Street Press.

Dobzhansky, Theodosius, 1967. *The Biology of Ultimate Concern.* New York: New American Library.

Dossey, Larry, 2014, *One Mind: How Our Individual Mind Is Part of a Greater Consciousness and Why It Matters,* Carlsbad, California: Hay House, Carlsbad, California.

Dyson, Freeman, 2015. *Dreams of Earth and Sky.* New York: New York Review Books.

Dyson, Freeman, 1988. *Infinite in All Directions. Gifford Lectures Given at Aberdeen, Scotland, April-November 1985,* New York: Harper and Row.

Echa, M. V., "On the Life and Legacy of Niels Bohr," in *Echa and Science! The Official Website for Post-modern Physics.*

Elgin, Duane 2018." Humanity's Journey Home: We Are Bio-Cosmic Beings Learning to Live in a Living Universe." Retrieved from http://newstoryhub.com/2018/02/humanitys-journey-home-we-are-bio-cosmic-beings-learning-to-live-in-a-living-universe1-by-duane-elgin/

Elgin, Duane, 2011. Retrieved from his website, http://truthseekers.cultureunplugged.com/truth_seekers/2011/12/our-living-universe.html. elgin 2011 "Our Living."

Elgin 2009 Quoted from an interview of Duane Elgin by Deepak Chopra, entitled "The Living Universe" on May 9, 2009, on Sirius XM Radio, as reported in the Huffington Post. Retrieved from You Tube https://video.search.yahoo.com/yhs/search?fr=yhs-itm-001&hsimp=yhs-001&hspart=itm&p=the+living+universe+chopra#id=2&vid=3c47c2198d27628e4b4f3c4f8248ae17&action=click

Elgin, Duane, December, 2000-February, 2001. "Our Living Universe." *Ions Review #54.* Institute of Noetic Sciences.

Elgin, Duane, 1993. *Awakening Earth: Exploring the Evolution of Human Culture and Consciousness.* New York: William Morrow.

Ellis, George F. R., December, 2015. "Does the Multiverse Really Exist? Proof of Parallel Universes Radically Different from Our Own May Still Lie beyond the Domain of Science. *Scientific American.*

Ellis, George F. R., 1993. *Before the Beginning: Cosmology Explained.* London, Great Britain: Bowerdean/Marion Boyars.

Falk, Dan, May 16, 2016. "New Support for an Alternative Quantum View." *Quanta Magazine.*

Fields, Helen, October 2010. "The Origins of Life: A Mineralogist Believes He's Discovered How Life's Early Building Blocks Connected Four Billion Years Ago." *Smithsonian Magazine*, October 2010.

Flatow, Ira, 2014. "New Research: Plant Intelligence May Forever Change HowYou think about Plants." Science Friday interview with Michael Pollan *Science Friday,* (hosted and produced by Ira Flatow), (https://www.pri.org/stories/2014-01-09/new-research-plant-intelligence-may-forever-change-how-you-think-about-plants.

Frankenberry, Nancy K., ed., 2008, *The Faith of Scientists: In Their Own Words.* Princeton University Press, Princeton, New Jersey,

Freeman, Morgan, (producer and host), 2010-2017. *Through the Wormhole.* PBS Documentary.

Gearhart, Mirjana, October-December 1979. "FORUM: From the Big Bang to the Big Crunch," North American AstroPhysical Observatory; *Cosmic Search*, Volume 1, Number 4.

Gefter, Amanda, April 4, 2016. "The Case Against Reality: Cognitive Scientist Donald Hoffman Explains Why Human Perceptions of an Independent Reality Are All Illusions," *The Atlantic.*

Gefter, Amanda, June 4, 2015. "A Private View of Quantum Reality: Quantum Theorist Christopher Fuchs Explains How to Solve the Paradoxes of Quantum Mechanics. His Price: Physics Gets Personal." *Quanta Magazine.* Retrieved from https://www.quantamagazine.org/20150604-quantum-bayesianism-qbism/.

Gefter, Amanda, June 30, 2014. "The Evolutionary Argument Against Reality." *Wired.* Retrieved from https://www.wired.com/2014/06/the-new-quantum-reality.

Gohd, Chelsea, December 2017. "This Discovery Could Change the Way We Search for Alien Life." *Futurism.*

Goldfarb, Ben, January 25, 2016. "The Forever War." *High Country News.*

Goodwin, Brian 1994. *How the Leopard Changes its Spots.* London: Weidenfeld and Nicholson.

Gott, J. Richard, and Vanderbei, Robert J., 2010. *Sizing Up the Universe: The Cosmos in Perspective.* Washington, D.C., National Geographic.

Gotswami, Amit; Reed, Richard E., and Gotswami, Maggie 1993. *The Self-Aware Universe: How Consciousness Creates the Material World.* New York, Jeremy Tarcher/Putnam Books.

Gould, Stephen Jay, 2009. *Stephen Jay Gould and the Politics of Evolution.* Prometheus Books, Amherst, New York.

Graves, Clare W., (Author) and Cowen, Chris, and Todorovic, Natasha (eds.), 2005. "The Never-Ending Quest: Dr. Clare W. Graves Explores Human Nature. Santa Barbara CA: ECLET Publishing.

Greene, Brian, (2004). *The Fabric of the Cosmos.* New York: Vintage Books.

Greene, Michael, November-December 2016. "Is VR a Game Changer: Virtual Reality in Therapy?" by Michael Washington, D.C., *Psychotherapy Networker.*

Greenstein, George, 1988. *The Symbiotic Universe: Life and Mind in the Cosmos.* New York: Morrow.

Harrison Edward R., 1985. *Masks of the Universe.* Cambridge, Great Britain: Cambridge University Press.

Hartle, J. B. 2005. "The Physics of Now." *American Journal of Physics.* Retrieved from http://arxiv.org/abs/gr–qc/0403001.

Harvey, Graham, 2006. *Animism: Respecting the Living World,* New York: Columbia University Press.

Hawking, Stephen, January, 21, 2013. "Stephen Hawking Quotes." Retrieved from http://www.educatinghumanity.com/2013/01/Stephen-Hawking-Quotes.html.

Hawking, Stephen, and Mlodinow, Leonard, 2010. *The Grand Design*. New York: Bantam.

Hawking, Stephen, and Mlodinow, Leonard, 2005. *A Briefer History of Time*. New York: Bantam.

Hawking, Stephen, 1988. A Brief History of Time. New York: Bantam.

Hawkins, David, 2014. *Power and Force*. Carlsbad, CA: Hay House.,

Heinlein, Robert A., 1961, *Stranger in a Strange Land. New York:* Putnam.

Hendrix, Harville, as referenced in de Quincey, Christian, 2005. *Radical Knowing*. Rochester VT: Park Street Press.

Herbert, Nick, "Elemental Mind: Human Consciousness in the New Physics," in Harvey Graham, 2006. *Animism: Respecting the Living World.* New York: Norton.

Heeren, Fred, 1995. *Show Me God.* Miamitown OH: Day Star Publications.

Hickman David, (series producer), McMaster Joseph (Writer, Producer, Director), *(2003). The Elegant Universe: Einstein's Dream* [TV series}. Boston: WGBH Educational Foundation, PBS NOVA. (Based on the book *The Elegant Universe,* by Brian Greene).

Hinshaw, Robert E., 2006. *Living with Nature's Extremes: The Life of Gilbert Fowler White*. Boulder CO: Johnson Books.

Hoffman, Donald, the YouTube video, *Simulation 2017: We Are Waking Up*. Retrieved from https://www.youtube.com/watch?v=fjbwWivYabU

Howes, Ryan, November/December 2016. "Food and Mood: What Every Therapist Needs to Know about Nutrition: An Interview with Joan Borysenko". *Psychotherapy Networker.*

Hoyle, Fred, 1982. *The Intelligent Universe.* New York: Holt, Rinehart and Winton.

Ingraham, Christopher, October 6, 2017. "We Have a Pretty Good Idea When Humans Will Go Extinct." *Washington Post.* Retrieved from https://www.washingtonpost.com/news/wonk/wp/2017/10/06/we-have-a-pretty-good-idea-of-when-humans-will-go-extinct/.

Jantsch, Erich, 1980. *The Self-Organizing Universe.* New York: Permagon Press.

Jeans, Sir James, 2010. *The Mysterious Universe.* Whitefish, MT: Kessinger Publishing.

Jones, Marina, February 13, 2014. "John Wheeler's Participatory Universe" *Futurism.* Retrieved from https://futurism.com/john-wheelersparticipatory-universe.

Kandel, Eric R., 2006. *In Search of Memory: The Emergence of a New Science of Mind New York:* W. W. Norton & Company.

Kant, Immanuel, 2002. "Critique of Pure Reason'" In M. C. Beardsley (ed.), *The European Philosophers from Descartes to Nietsche.* New York: The Modern Library

Kauffman, Stuart A., 2008. *Reinventing the Sacred: A New View of Science, Reason, and Religion.* New York: Basic Books.

Kauffman, Stuart A., November 21, 2007 "Beyond Reductionism: Reinventing the Sacred". *Zygon Journal of Religion and Science.* Retrieved from https://doi.org/10.1111/j.1467-9744.2007.00879.x

King, Barbara J., January 12, 2017. "How Smart Are Horses?" *NPR.* Retrieved from http://www.npr.org/sections/13.7/2017/01/12/509451392/how-smart-are-horses.

Kitchen, William H, 2017. *Philosophical Reflections on Neuroscience and Education. London:* Bloomsbury Publishing.

————"Know Your Nevada Neighbors" Retrieved from http://www.knowyourneighbors.net

Lane, Nick. 2015. *The Vital Question: Energy, Evolution, and the Origins of Complex Life. New York:* W.W. Norton and Company Inc. First published in Great Britain under the title, *Why Is Life the Way it is?*

Lanza, Robert. www.robertlanza.com. Retrieved from http://www.robertlanza.com/why-are-you-here-a-new-theory-may-hold-the-missing-piece /.

Lanza, Robert, and Berman, Bob, 2009. *Biocentrism: How Life and Consciousness are the Keys to Understanding the True Nature of the Universe.* New York: BenBella Books.

László, Ervin, 2007. *Science and the Akashic Field: An Integral Theory of Everything.* Rochester, VT: Inner Traditions, Inc.

Laszlo, Pierre, 1997. *Origine de la Vie:* 100,000 Milliards de Scénarios. Paris: La Recherche

Leibnitz, Godfried, 1670. *Meditations on Knowledge, Truth, and Ideas;* the *Discourse on Metaphysics.* Public domain.

Lewontin, Richard C., 1991, *Biology as Ideology: The Doctrine of DNA*, New York: Harper Perennial, as quoted in Prindle, David F.,2009, *Stephen Jay Gould and the Politics of Evolution*. Amherst, New York: Prometheus Books.

Lipton, Bruce H., and Bhaerman, Steve, 2009. *Spontaneous Evolution: Our Positive Future and a Way to Get There From Here*. Carlsbad CA: Hay House.

Lipton, Bruce, 2005, *The Biology of Belief, Unleashing The Power Of Consciousness, Matter And Miracles*. Carlsbad, CA: Hay House.

Lloyd, Seth, August 27, 2007. *Life: What a Concept!* An Edge Special Event at Eastover Farm: A conversation with Seth Lloyd." Retrieved from https://www.edge.org/conversation/seth_lloyd-seth-lloyd%E2%80%94life-what-a-concept.

Lloyd, Seth, 2006, *Programming the Universe: A Quantum Computer Scientist Takes On the Cosmos*. New York Alfred A. Knopf.

Lovelock, James, 2000. *Gaia: A New Look at Life on Earth*. Oxford Great Britain, Oxford University Press: Oxford Great Britain.

Lumière, Auguste, and Lumière, Louis, (directors), 1895. *Arrival of a Train at La Ciotat*. Original Title: *L'Arrivée d'un Train Ã la Ciotat*. Retrieved from https://www.youtube.com/watch?v=1dgLEDdFddk

Margenau, Henry and Varghese, Roy A., eds., 1992. *Cosmos, Bios, Theos: Scientists Reflect on Science, God, and the Origins of the Universe, Life, and Homo Sapiens*. Chicago: Open Court Publishing.

Marshall, Michael, August 22, 2012. "DNA Could Have Existed Long before Life Itself." *The New Scientist*.

Martin, Sean, September 5th, 2018. "Humans WILL Go Extinct and There's NO Hope for Survival: It's WHEN Not If, Biologists Say." UK's *Express. UK's Express.* Retrieved from https://www.express.co.uk/ news/ science/1013466/human-mass-extinction-evolution-university-ofKansas.

Mayr, Ernst F. 1982. *The Growth of Biological Thought: Diversity, Evolution, and Intelligence.* Cambridge MA: Bellnap Press, Harvard University.

McIntosh, Steve, 2015. *The Presence of the Infinite: The Spiritual Experience of Beauty, Truth, and Goodness.* Wheaton IL: Quest Books.

McIntosh Steve, 2012. *Evolution's Purpose: An Integral Interpretation of the Scientific Story of Our Origins. New York.* Select Books.

McIntosh, Steve, 2007. *Integral Consciousness and the Future of Evolution,* New York Paragon.

McMaster, Joseph, Judd Graham, and Streeter, Sabin, (writers), Paula S. Apsell, (senior producer), Melanie Wallace, (senior series producer), Brian Greene, (executive ed.) 2011. "*Fabric of the Cosmos.* PBS NOVA Production.

McTaggart, Lynne, 2008. *The Field: The Quest for the Secret Force of the Universe.* New York: Harper.

Morell, Virginia, September 21, 2016. "Horses Can Use Symbols to Talk to Us." *Science Magazine.* http://www.sciencemag.org/ news/2016/09/horses-can-use-symbols-talk-us.

Morris, Errol, (director) 1991. *A Brief History of Time* documentary. Norwich, Great Britain: Anglia Television and Tokyo Broadcasting.

Moseman, Andrew, December 1, 2010. "The Estimated Number of Stars in the Universe Just Tripled." *Discover Magazine*. Retrieved from http://blogs.discovermagazine.com/80beats/2010/12/01/the-estimated-number-of-stars-in-the-universe-just-tripled/#.XUYwq-hKh5l.

Moskowitz, Clara, 2010. "Life's Great Mystery. What Exactly Is Life?" *LiveScience*. Retrieved from http://www.nbcnews.com/id/40630735/ns/technology_and_science-science/t/lifes-great-mystery-what-exactly-life/#.XUYZ9ehKh5k.

Naess, Arne, (2005-7). *Morning Earth,* "An Introduction to Deep Ecology." *Morning Earth*. Retrieved from http://www.morning-earth.org/deep-ecology.htm.

Narby, Jeremy, 2006. *Intelligence in Nature: An Inquiry into Knowledge.* New York: Jeremy P Tarcher.

Narby, Jeremy, 1998. *The Cosmic Serpent: DNA and the Origins of Knowledge.* New York: Jeremy P. Tarcher.

North American AstroPhysical Observatory, (Fall) October-December, 1979. "FORUM: From the Big Bang to the Big Crunch," *Cosmic Search*, Volume 1 Number 4.

Nash, Madeline J., December 5, 1995. "When Life Exploded," *Time Magazine*.

Paulson, Steve, May 4,2017. "Roger Penrose On Why Consciousness Does Not Compute: The Emperor of Physics Defends His Controversial Theory of Mind." *Nautilus (Online Magazine)*.

Peat, F. David, 1997. *Infinite Potential: The Life and Times of David Bohm.* New York: Basic Books.

Penrose, Roger, 1991. As quoted in *A Brief History of Time*, a biographical documentary film about the physicist Stephen Hawking, directed by Errol Morris.

Planck, Max, 1944. *Das Wesen der Materie* [*The Nature of Matter*], a speech delivered by Planck in Florence, Italy.

Pollack Robert, February 1,1997. "A Crisis of Scientific Morale," *Nature*. 385: 673-4

Pollack, Robert, 1994. *Signs of Life: The Language and Meanings of DNA*. New York: Viking.

Pollan, Michael, December 23, 2013. "The Intelligent Plant: Scientists debate a new way of understanding flora," *Science Friday.*_Michael Pollan, author of *The Botany of Desire, The Omnivore's Dilemma,* and others, discusses with *Science Friday,* (hosted and produced by Ira Flatow), his 2013 article in the *New York Times*, entitled "The Intelligent Plant: Scientists debate a new way of understanding flora," http://www.newyorker.com/magazine/2013/12/23/the-intelligent-plant.

Popper, Karl, R., 1990. *A World of Propensities,* Bristol, England: Thoemmes.

Popper, Karl R., 1974. *Unended Quest: An Intellectual Autobiography,* London: Fontana.

Prasad, (ed), 1987. "Our Home in Space.' In R.C. Prasad (ed.), *Modern Essays: Studying Language through Literature.* New Delhi, India: Vikas Publishing House Private Limited.

Pritchett Laura, November 9, 2015. "Back to the Basics: An Ode to Germs, Guts, and Gardens." *High Country News,* Vol. 47, No.19.

Quantum Physics at Carnegie-Mellon University webpage. According to the Quantum Physics at Carnegie-Mellon University webpage, retrieved from http://quantum.phys.cmu.edu/CHS/histories.html.

Rae, Alastair I. M., 2005. *Quantum Physics: A Beginner's Guide.* London: Oneworld Publications.

Rao, K. Ramakrishna, 2011. "Cognitive Anomalies, Consciousness, and Yoga, Volume XVI, Part 1." *History of Science, Philosophy, and Culture in Indian Civilization,* (D.P. Chattopadhyaya, general editor.) New Delhi, India: Centre for Studies in Civilization and Matrix Publishers (joint publishers).

Redd, Nola Taylor, March 14, 2018. "Stephen Hawking Biography1942-2018." *Space.com.* Retrieved from https://www.space.com/15923-stephen-hawking.html.

Redd, Nola Taylor, April 8, 2015. "Basic Ingredients for Life Found around Distant Star." *Space.com.* Retrieved from www.space.com/29049-life-ingredients-found-around-star.html.

Ringohofer, Monamie, and Yamamoto, Shinya, (Kobe University). December 15, 2016. "When Horses Are in Trouble, They Ask Humans for Help." *Animal Cognition.* Retrieved from http://www.kobe-u.ac.jp/en/NEWS/research/2016_12_15_01.html.

RMPBS PBS Library. Retrieved from http://www.pbs.org/wgbh/evolution/library/03/5/l_035_01.html.

Robsville, Sean. Retrieved from kwelos.tripod.com/emergent.htm.

Rolston III, Holmes, 2010. *Three Big Bangs: Matter-Energy, Life, Mind.* New York: Columbia University Press.

Rolston III, Holmes, 2006. *Science and Religion: A Critical Survey Reprinted with a New Introduction.* Philadelphia: Templeton Foundation Press.

Rolston III, Holmes, 1999. *Genes, Genesis and God,* Cambridge United Kingdom: Cambridge University Press.

Roman Inquisition, 1615. Public domain.

Rosenblum, Bruce, and Kuttner, Fred, 2011. *Quantum Enigma: Physics Encounters Consciousness.* Oxford, United Kingdom: Oxford University Press.

Safina, Carl, 2015. *Beyond Words: What Animals Think and Feel.* New York: Henry Holt and Company, LLC.

Sahtouris, Elisabet, November 10, 2017. *Beyond Darwin: Sahtouris, Science and Hope (Hawaii Is My Main Land).* ThinkTech Hawaii. Retrieved as https://www.youtube.com/watch?v=pFmTKZkdcXI.

Sahtouris, Elisabet, 2007. "Chapter 7 Prologue to a New Model of a Living Universe." *Mind Before Matter: Vision of a New Science of Consciousness.* New York: Iff Books.

Sahtouris, Elisabet, 2000. *EarthDance: Living Systems in Evolution.* Lincoln NE: iUniversity Press.

Sahtouris, Elisabet, *Darwin Part 1*, a video conversation on her website, www.Sahtouris.com.

Schlosshauer, Maximilian, Kofler, J, and Zeilinger, A., 2011. *A Snapshot of Foundational Attitudes Toward Quantum Mechanics.* Retrieved from http://arxiv.org/pdf/1301.1069v1.pdf).

Schrödinger, Edwin, 1995. *The Interpretation of Quantum Mechanics: Dublin Seminars (1949-1955 And Other Unpublished Essays).* Barnsley, United Kingdom: Ox Bow Publishers.

Scoles, Sarah, 4/19/2016. "Think Big: Can Physicists Ever Prove the Multiverse Is Real? Astronomers Are Arguing about Whether They Can Trust This Untested—and Potentially Untestable—Idea." *Smithsonian.com.* Retrieved from https://public.media.smithsonianmag.com/filer/24/1a/241a97c2-711a-4c19-8b55-4c0b53a0618b/10722-thinkbig.png.

Siegfried, Tom, 2014. "Tom's Top 10 Interpretations of Quantum Mechanics." *Science News*. Retrieved from https://www.sciencenews.org/blog/context/toms-top-10-interpretations-quantum-mechanics

Siegfried Tom, January 21, 2014. "Gell-Mann, Hartle Spin a Quantum Narrative about Reality." *Science News*. Retrieved from https://www.sciencenews.org/blog/context/gell-mann-hartle-spin-quantum-narrative-about-reality.

Semeniul, Ivan, January 31, 2017. "195-Million-Year-Old Dinosaur Bone Yields tTaces of Soft Tissue." *The Globe and Mail*. Retrieved from http://www.theglobeandmail.com/technology/science/195-million-year-old-dinosaur-bone-yield-traces-of-soft-tissue/article33846083/

Shapiro, James, 2011. *Evolution: A View from the 21ˢᵗ Century."* Upper Saddle River, NJ: FT Press Science.

Shapiro, James, 1996. *Origins: Skeptics Guide to the Creation of Life on Earth*. London: Heinemann.

Sheldrake, Rupert, 2012. Science Set Free: 10 Paths to New Discovery. New York: Deepak Chopra (Publisher).

Silver, Joel, (Producer), Wachowski, Lilly, and Wachowski, Lana, (Directors}, 1999.*The Matrix* (Motion picture). USA: Warner Brothers Pictures.

Spinoza, Benedict de, 1677. *The Ethics - Ethica Ordine Geometrico Demonstrata* (translated from the Latin by R. H. M. Elwes), 1677. (Amazon): CreateSpace Independent Publishing Platform.

Stapp, Henry, P., 2001. "Quantum Theory and Quantum Theory and the Role of Mind in Nature." Berkeley CA: Lawrence Berkeley National Laboratory University of California. Retrieved from http://wwwphysics.lbl.gov/~stapp/vnr.pdf .

Stapp, Henry, P., 2001. "Von Neumann's Formulation of Quantum Theory and the Role of Mind in Nature. "Berkeley CA: Lawrence Berkeley National Laboratory University of California. Retrieved from https://arxiv.org/pdf/quant-ph/0101118.pdf.

Stanford Encyclopedia of Philosophy 1995. The Stanford Encyclopedia of Philosophy (1995) is an online encyclopedia of philosophy with peer-reviewed publication of original papers in philosophy, maintained by Stanford University.

Stegen, Anne, March 17, 2017. "Amazing: Bands of Wild Horses Line Up to 'Say Goodbye' to Mare Who Died." KPNX 12:12 PM., reporting on the work of the Salt River Wild Horse Management Group, outside Phoenix, Arizona.

Steinfeld, Alan, 2012. "An Interview with Deepak Chopra. *New Realities,* www.newrealities.com.

Stewart, John, 2000. *Evolution's Arrow: The Direction of Evolution and the Future of Humanity.* Orange, CA: Chapman Press.

Storoy, David, August, 2014. "David Bohm, Implicate Order and Holomovement" Retrieved from https://www.scienceandnonduality.com/david-bohm-implicate-order-and-holomovement.

Strotz, Luke, August 22, 2018. What Makes Some Species More Likely to Go Extinct?" *Smithsonian Magazine.* Retrieved from https://www.smithsonianmag.com/science-nature/what-makes-some-species-more-likely-go-extinct-180970103/#g22TV0jcA7xDGSFD.99.

Sullivan, J. W. N. January 25th, 1931. "Interviews With Great Scientists—Max Planck." *The Observer.*

Talbot, Michael, 1991. *The Holographic Universe.* New York: Harper-Collins.

Teilhard de Chardin, Pierre 2008, *The Phenomenon of Man.* New York: Harper Perennial Modern Classics.

Teilhard de Chardin, Pierre, 2004 *The Future of Man,* New York: Image Books.

Vallary, Anna. "Farm Animals that Are Probably Smarter Than Your Dog." One Green Planet. Retrieved from http://www.onegreenplanet.org/animals ansdnature/farm-animals-that-are-probably-smarter-than-your-dog/.

Virtual Reality Society 2017, "What Is Virtual Reality?" Retrieved from https://www.vrs.org.uk/virtual-reality/what-is-virtual-reality.html.

Veissiére, Samuel, August 17, 2017. "When Healing Is a No-Brainer." *Psychology Today.*

Velmans, Max, 2000. *Understanding Consciousness.* London, Routledge.

Vertosick, Frank, 2002. *The Genius Within.* New York: Harcourt.

Von Braun, Wernher, and Powell-Willhite, Irene, (ed.), 2007. *The Voice of Dr. Wernher Von Braun: An Anthology.* Canada: Apogee Books.

Walia, Arjun, March 8, 2014. Retrieved from "Science Proves That Human Consciousness and Our Material World Are Intertwined: See For Yourself." Collective Evolution.com. Retrieved from https://www.collective-evolution.com/2014/03/08/10-scientific-studies-that-prove-consciousness-can-alter-our-physical-material-world/.

Wallace, B. Alan, 2008. *Embracing Mind: The Common Ground of Science and Spirituality.* Boulder, CO: Shambhala.

Walker, Evan Harris, 2000. *The Physics of Consciousness. The Physics Of Consciousness: The Quantum Mind And The Meaning Of Life.* New York: Basic Books.

Watson James D., 1987, *Molecular Biology of the Gene.* Menlo Park, CA: Benjamin/Cummings Publishing.

Watson, James D., & Crick, Francis, H. C., 1953. "A Structure for Deoxyribose Nucleic Acid. *Nature* 171, 737–738.

Wesson, Robert, 1991. *Beyond Natural Selection,* Cambridge, Massachusetts: MIT Press.

Whitesides, George M., 2008. "Forward: The Improbability of Life." In John D. Barrow, Simon Conway Morris, Stephen J. Freeland, and Charles L. Harper Jr., eds. *Fitness of the Cosmos for Life: Biochemistry and Fine Tuning.* Cambridge: Cambridge University Press.

Whitworth, Brian. *Quantum Realism,* 2014. An online paper, retrieved from http://brianwhitworth.com/BW-VRT1.pdf f.

Wigner, Eugene P., 1960. "The Unreasonable Effects of Mathematics in the Natural Sciences." *Communications on Pure and Applied Mathematics.* A monthly scientific journal, published by John Wiley and Sons.

Wilber, Ken, 2000. *A Theory of Everything: An Integral Vision for Business, Politics, Science and Spirituality,* Boulder, CO: Shambhala.

Wilson, Edward O., and. Kellert, Stephen R., 1993. "Biophilia and the Conservation Ethic," in *The Biophilia Hypothesis.* Washington, D.C., Shearwater Books.

Wilson, Edward O., 1992. *The Diversity of Life.* New York: Penguin.

Wolchover, Natalie (6/30/14). Have We Been Interpreting Quantum Mechanics Wrong this Whole Time? *Wired,* Retrieved from https://www.wired.com/2014/06/the-new-quantum-reality/.

Wolfson, Richard, (Great Courses lecture), from the course "Einstein's Relativity and the Quantum Revolution."

Yeats, David A., 2014. *Co-Creating a Brilliant Relationship: A Journey of Deepening Connection, Meaning, and Joy.* Parker, CO: Outskirts Press.

Yeats, Julian, 2018. *Everything You Need to Know.* (Unpublished from the album tentatively titled "Julian Yeats and the Dream Doctors: Vibe").

Yeats, Julian, 2018. *Higher Vibe.* (Unpublished from the album tentatively titled "Julian Yeats and the Dream Doctors: Vibe").

Yeats, Julian, 2018. *Pass the Joy.* (Unpublished from the album tentatively titled "Julian Yeats and the Dream Doctors: Vibe").

Yeats, Julian, 2018 *Timeless Being.* (Unpublished from the album tentatively titled "Julian Yeats and the Dream Doctors: Vibe").

Yost, Peter, (writer, director, and producer), 2015. *Invisible Universe Revealed.* Boston: WGBH: Pangloss Films, 2015).

Young, Jon, 2012. *What the Robin Knows: How Birds Reveal the Secrets of the Natural World.* Boston: Houghton Mifflin.

Zeh H. Dieter, 2004. "The Wave Function: It or Bit?" p. 4. In J. D. Barrow, P. C. W. Davies, & J. Charles L. Harper (Eds.), *Science and Ultimate Reality: Quantum Theory, Cosmology and Complexity.* Cambridge: Cambridge University Press. Retrieved from http://arxiv.org/abs/quant–ph/0204088.

Zehavi, Idit, and A. Dekel, Avishai, 1999. "Evidence for a Positive Cosmological Constant from Flows of Galaxies and Distant Supernovae." *Nature* 401: 252-254.

Zohar, Dinah, 1991. *The Quantum Self.* London: Flamingo.

INDEX

Combs, Allan, 59, 60, 190
community, multicellular, 179
complementarity, 21, 42, 47, 56, 79
complementary change, 46
complexification, 240, 276
complexifies, 57
complex interactions, 163
complexity, 12, 26, 43, 52, 86, 95
    increasing, 169, 244
    organizational, 240
complex life, 161, 301
connectedness, universal, 44, 56
conscious decision, 124
consciousness
    collective, 262
    definition of, 55
    existence of, 89, 96
    expansive, 141
    observer's, 49
    primacy of, 87
Consciousness Paradigm, ix, 17, 29
    unitary, 231
conscious observer-participant, 53
Consistent/Decoherent Histories, 82
Consistent/Decoherent Histories inter-
    pretations, 83
Consistent Histories, 72, 82
continuity, 32, 139
Conway, John, 55
Cook-Greuter, Susanne, v, 260
Cook-Greuter's Developmental Models,
    260, 294
Cook-Greuter's Self-Identity Model, 259
Copenhagen, 68–71, 73–74, 82
Copenhagen approach, 83
Copenhagen interpretation, xiii, 40, 43-
    43, 45, 48, 51, 59, 68–74, 82, 83
Copernicus, 33, 35–37
Corning, Peter, 229, 294
cosmic nucleosynthesis, 12
cosmology, inflationary, 12
Cowan, 258, 260, 292
Crease, Robert, 76
Crew, Bec, 61
crew member, 252
Crick, Francis, 157-158, 222–23, 294,
    311
crop pollinators, 186

crow family relatives, 188
crows, 188–89, 198
Curie, Marie, 39

D

Damasio, Antonio, 213
Dark Ages, 33
Das, Dhruba, 196
Dass, Ram, vi
Darwin, Charles, xviii, 63-64, 95, 106,
    159-160, 226-227, 230, 232, 234.
da Vinci, Leonardo, 175
Davies, Paul, 8, 13, 47
DBB, 80
de Broglie, Louis, 40
de Broglie/Bohm interpretation, 45, 80
decoherence, 41, 83
Decoherent Histories interpretations,
    82–84
De Duve, Christian, 161, 247, 294
Dekel, Avishai, 13
Denton, Michael, 170
Denver, John, 246
de Quincey, Christian, xiv, xx, 18,
    20–22, 25, 27–28, 31, 32, 46–49,
    52–54, 62, 71, 89, 93–94, 96–97,
    99–100, 216, 238
Descartes, René, 35–38, 87, 97
Descartes' dualism, 85
determinism, 27, 30–31, 39
determinist, 38
deterministic chain, 27
development
    evolutionary, 156, 257, 268
    spiral of, 244, 261
developmental levels, 89, 156
device, measuring, 85
de Waal, Frans,190, 198
dimensions, multiple, 60
Dion, 105
Dirac, Paul, 40
distances, super-luminal, 47
DNA, 23, 28
    essence and origins of, 159–60
DNA and RNA molecules, 173, 188,
    224
DNA letters, 176
DNA nucleotides, 158

316

DNA's role, 223
DNA's species, 177
Dobzhansky, Theodosius, 102, 295
dolphins, 171, 188, 197–98, 203, 205–7
Dossey, Larry, 126, 146
dualism, 87, 97, 183
Dualists, 104
Dyer, v
Dyson, Freeman, xx, 48, 54-55, 169,
    182-183, 243,253, 281-282

E
Early Modern Period, 35
earth, primordial, 158
Earthlife, 172, 175, 180
Earth's evolution, 177
Earth's surface, 164, 168
Echa, M. V., 66
ECLET ("Emergent Cyclical Levels of
    Existence Theory"), 258
ecosystems, composite, 175
Eddington, Arthur, 12, 126, 129, 247
Eckhart, Meister, vi
eggs, wasp's, 185
Einstein, Albert, viii, xv, 12, 18, 22, 25,
    30–31, 33, 38–39, 43, 46, 51–53,
    55, 60, 62, 70, 80–82, 146
Eldredge, Niles, 226
electrical charge, 11
electrons, 19, 24, 28, 39–40, 42, 44–45,
    48–49, 51, 54–55, 75–76, 99
    atom's, 11
electron waves, 40
elephant cerebellum, 187
Elgin, 19–20, 99, 155, 169
Ellis, George, 12, 77
emergence, 40, 55, 64, 90, 96
    adaptive, 227
    informed, 160
    synchronous, 171
Emergent Movement, 261
EM-field, universal, 24
empathy, 195, 204
energy
    dark, 43, 91
    electromagnetic, 39
    packets of, 19, 22, 31
    vibrating, 18, 20

energy reactions, 162
energy sea, 16
Enlightenment, 35, 37
entanglement, 42, 44, 47, 56, 79, 81
entities, elementary, 46
environment, 36, 44, 56, 83, 163
environmentalism, 255
epigeneticists, 102
epigenetic research, 102
epigenetics, 26
epiphenomena, 38
EPR controversy, 81
equilibrium, 151–52
    punctuated, 226–27
Erickson, Erik, 260
ether, 24
    luminiferous, 24
    physical, 134
ethologists, 183, 198
European Renaissance, 33
events
    deterministic, 27
    random, 119, 126
    self-caused, 47
    uncaused, 47
Everett, Hugh, 76
Everett interpretation, 74
evolution, 17, 28, 32, 39, 55, 64, 90
    interior, 257
    prebiotic, 159
    progressive, 280
    the evolutionary argument, 53
evolutionary awareness, 180, 279–81
evolutionary development compounds,
    179
evolutionary objectives, 281
evolutionary process, 57
    ongoing, 180, 279
evolutionary processes, on-going, 280
evolutionary theory, 221, 232, 234
Evolution's Arrow, 279
Executive Movement, 261
expansion, 10, 64, 75
experience
    immediate, 122
    occasions of, 94
experiments
    double slit, 39

245

319

von-Neumann-Wigner, 84
intersubjective, 73
intersubjectivity, 53
invertebrates, 188
Ithaca interpretation, 86

J

Jantsch, Erich, 244-245
Jeans, James, 88
Jesus, v
John of the Cross, v
Jones, Marina, 69

K

Kandel, Eric R., 105–8,
Kant, Immanuel, 50
Karban, Rick, 212
Kauffman, Stuart, xviii-xix, 43, 55, 90, 228
Keepin, Bill, 20
Kegan, Paul, 134, 292
Kepler, Johannes, 33, 35–37
Kepler, Katharina, 36
King, Barbara, 193
Kirchhoff, Gustav39
Kistiakowsky, Vera, 13
Kitchen, William H., 54
knowledge, shamanic, 210
Koch, Christof, 182
Kochen, Simon, 55
Koestler, Arthur, 268
Kofler, Johannes, 72
Kohlberg, Lawrence, 260
Krishnamurti, Jiddu, 60
Kurzweil, Ray, 170

L

Lamarck, Jean-Baptiste, 222
landscape, evolutionary, 226
Lane, Nick, 117, 161–64, 301
Lanza, Robert, 11, 20–21, 27, 42, 49–51, 53, 59, 71, 74, 81, 91–92, 96, 98–99, 107, 114, 117–18, 121, 124–25, 142–144, 146, 185, 251
Laplace, Pierre-Simon, 27
László, Ervin, 15–17, 24–25, 46–48, 56, 170, 232, 247, 301
Laszlo, Pierre, 232

Late Middle Ages, 33
law of complexity-consciousness, 240
learning
  human speech, 193
  song, 193
Leibniz, Gottfried, 63
lens, subjective, 74
Levy, Matthew, 158
Lewontin, Richard, 102
Libet, Benjamin, 124, 185
life and consciousness, 11, 96, 98–99, 142, 247, 274–75
life conditions, 256–58, 261, 277
life forms, 58, 89
lifeforms, simple multicellular, 182
life forms, single, 179
Linnaeus, Carl, 222
Lipton, Bruce, iv, 25, 30–31, 56, 180, 188, 217
Lipton and Bhaerman, 26, 30–31
living evolutionary unfolding, 180
living systems, healthy, 156, 167, 170
living systems theory, 21
Lloyd, Seth, 113–14, 127, 137, 153–55, 160–61, 165,
Lockean empiricism, 183
Lorimer Moseley, 273
Lovelock, James, 176–77, 247, 302
Lumière Brothers, Auguste and Louis, 122
Luther, Martin, 34, 36

M

Macintyre, Rick, 195
Mach, Ernst, xv
macroevolution, 230
macromolecules, 162
mammalian species, 289
mammals, herding, 192
manifests, 19–20, 24, 45, 55, 63, 98
Many Worlds Interpretation, (MWI), 74-76, 78–79
mares, young wild, 194
Margenau, Henry, 12–13
Marshall, Michael, 158-159, 302
Martin, Sean, 289, 303
material, inorganic, 168
materialism, scientific, 25, 43, 59, 93

materialist, 38
Materialistic Paradigm, 17, 27, 29
Materialist Paradigm, ix, xiii, 17, 27, 29
Materialist Paradigm perspective, 241
materialist philosophy, 27
material primacy, 120
material world, 32–33, 39, 66, 68, 97
mathematical equations, 40, 71
matter
  dark, 43, 91
  inorganic, 161
  new, 157
  subatomic, 22
  unconscious, 54
Maturana, Humberto 157
Maxwell, James Clerk, 39
Mayr, Ernst, 224-225, 303
McIntosh, Steve, xxi-xxii, 111, 159, 190, 228–31, 239–41, 243–45, 247–52, 255, 257–61, 263–69, 282–83, 303
McMaster, Joseph, 23, 62, 71
McTaggart, Lynne, 21–23
meaning, 23, 27, 34–35, 44, 55–56, 62, 79–81, 90, 94, 100
measurement/observation, 84
measurements, 39, 42, 48, 50, 67, 69, 73, 77, 82, 84
mechanics, solid, 39
mechanisms, evolutionary, 231
mechanistic verses quantum principles, 107
mechanomorphism, 207
Medieval Period, 33
memetic codes, 260, 292
metaphysics, 67, 69
metaverse, 14, 16–18, 48
metaverse references, 170
Michaelson-Morley experiments, 24
microbes, 178, 211
microevolution, 230
microscopic fungus, 178
Midas, 46
Middle Ages, 33
Miller, Christopher, 173,177, 224, 247
mind, nonmaterial, 38
mind-brain relationship, 109
mirror-as-reflector, 278
Mlodinow, Leonard, 52–53

modal interpretations, 86
Modern Age, 35
modernist consciousness, 255
Modern Period, 34
Mohammed, 5
molds, unicellular slime, 188
molecular biology, incorporated, 106
monism, 87
monkey, 186, 199, 220
Monte Vista National Wildlife Refuge, 191
Morell, Virginia, 194, 303
Morris, Errol, 12
Moseman, Andrew, 6
Moskowitz, Clara, 152, 304
Moss, Carrie-Anne, 26
Moss, Cynthia, 197
motion, 19, 22–23, 38, 50, 60, 69
movement, holographic, 239
multiverse, 74–77
  inflationary, 75
Musk, Elon, 121
mutations, 221, 223, 236

N
Naess, Arne, 64, 247
Nagel, Thomas, 94, 239
Nakagaki, Toshiyuki, 191
nanotechnology, 26, 40
nanotornados, 26
Narby, Jeremy, 55, 156, 173, 181, 191, 210, 221, 226, 229
NASA website, 151
Nash, Madeline, 225-226
native human consciousness, 180, 239, 241, 243, 245, 247, 249–53, 255, 257, 259, 261, 263, 265, 267, 269
natural realists, 104
natural selection, 64, 83
nature, intelligence in, 172–73, 175, 177, 179, 181, 183, 185, 187, 189, 191, 193, 195, 197, 199, 201, 203, 205, 207, 209, 211, 213, 215, 217
Neanderthal ancestors, 289
Neo-Darwinian biologists, 245
Neo-Darwinian evolution, xviii, 230–31
neo-Darwinians, 159, 221, 230
Neo-Darwinism, 160, 217, 223, 231

plenum, active, 25
Pollack, Robert, 177, 305
Pollan, Michael, 208–14, 247, 305
    speculates, 215
Pool, Joyce, 197
Popper, Karl, 79, 218, 227, 232,
populations, 33, 202, 230
    dispersed, 178
    single ancestral, 176
population's mutation, 151
positron-emission tomography (PET),
    106
postmodernists value multiculturalism,
    255
post-modern physics, 66
post-postmodernists, 255
post-reductionistic, 221
Powner, Matthew, 158–59
Szostak, Jack, 159
Pranzetti, Daniele, 61
Prasad, R. C., 7
prebiotic chemists, 158
precursors, 163, 173
predisposition, 195
Preparatory Movement, 261
Pribram, Karl, xiii, 59
primates, 187, 197, 204
Pritchett, Laura, 156
probabilities, 28, 41–43, 45, 49, 53, 57,
    70–71, 73, 78, 82–84
    mathematical, 48, 51
probability calculation, 41
probability rules, 46
probability waves, 32, 40, 44, 143
processes
    autopoietic, 218
    duplication, 222, 224
    self-maintaining, 218
progress, evolutionary, 245, 250
properties, emergent, 38
proteins, 158, 164, 168, 173–74, 222,
    224, 228
Protestants, 36
Puthoff, Hal, 22

Q
QBism, 45, 72–74
quanta, 22, 32, 42, 44–47, 100

quantized form, 39
Quantum Bayesianism, 73
quantum bonding, 40
quantum collapse, 69
quantum communication, 153
quantum computation, 153
quantum computer, 113–14, 153–54
    great, 154
quantum consciousness, 127
Quantum Darwinism, 83
quantum data, 45, 64, 78, 80, 85
quantum enigma, unsolved, 45
quantum entanglement, 43
quantum entity, 55, 57–58, 240
quantum events, 31–32, 47–49, 53, 57,
    82–83, 120, 140
    self-caused, 47
    uncaused, 47
quantum expansion, 55
quantum experiments, 52, 76, 85
quantum field, 22
quantum grid, 134
Quantum Information interpretation, 86
quantum interpretation, 80
quantum jumps, 24, 40
quantum level, implicate, 47
Quantum Logic interpretation, 86
quantum mechanics, 13, 21, 26, 31–32,
    38, 41, 45, 51, 53, 56, 67–68,
    70–74, 79, 81–82, 86, 99
    fundamental, 82
quantum/mind, xvi 100
Quantum Mind Paradigm, x, 145, 243,
    282
quantum nonlocality, 47
quantum observation, 48
quantum observer, xvi, 49
quantum paradox, 41
quantum physics, 20, 30, 40, 42–44,
    48–49, 57, 59, 68, 70, 81–84, 86
quantum potentialities, 92, 126
quantum probabilities, 49
    collapsing, 32
quantum randomness, 43
quantum realism, 21, 42, 45, 55
Quantum realism and virtual reality, 140
quantum reality, 21, 43, 46, 79
quantum relationships, 43

quantum rules, 70
quantum science, 45, 49, 63
quantum sea, 23, 25
quantum states, 26, 43, 54, 56–57, 73
quantum universe, 32, 44
quantum vacuum, 16
quantum wave functions, 112
quantum world, 25–26, 42, 51
quasiclassical domains, 83

R
radiation, black box, 39
Rae, Alastair, 40, 42, 51, 67, 70, 72–73, 76–78, 81
random changes, 223, 231
random genetic mutations, 221
random mutation, 159, 244
randomness, 63, 69
Rao, K. Ramakrishna, 126, 271, 306
rapid evolutionary change, 226
ravens, 189, 203, 207
reality
  conscious, 141
  emergent, 117
  game's, 141
  holodeck, 252
  illusory, 239
  non-physical, 239
  subjective, 62
Redd, Nola Taylor,165–66, 175, 306
reductionist worldview, 43, 55
Rees, Martin, 53, 71
Reformation, 35
Reiss, Diana, 206
Relational Quantum Mechanics. *See* RQM
relationships, 23, 34, 59, 70, 72–73, 79, 82, 84–85, 91
relativity, 12–13, 26, 40, 47, 53, 62
relativity theory, 21, 52, 62
representation of what is, 252, 286
reproduction, 151, 161, 169–70, 218
revolution, industrial, 40
Ringhofer, Monamie, and Yamamoto, Shinya, 195
RNA and DNA world, 159
RNA-based life, 158
RNA molecules, 173, 188, 224

RNA nucleotides, 158–59
RNA world, 158
Robsville, Sean, 101, 306
rock formations, 168
rocks, 38, 53, 62, 84, 89
  inanimate, 169
Rolston III, Holmes, iii, 11, 78, 103, 112, 114, 150, 151,156–57, 159, 162, 179, 181, 227, 247, 252
Roman Inquisition, 36
Röntgen, Wilhelm Conrad, 39
rooks, 188–89
Rosenblum Bruce, and Kuttner, Fred, 14, 26, 44–45, 56, 68–71, 83
Rothman, Tony, 12
RQM (Relational Quantum Mechanics), 45, 72–74
Runaround Sue, 105
Rutherford, Earnest, 39
Rydberg, Johannes, 39

S
Sacks, Oliver, 187
Safina, Carl, 182, 184–85, 187–89, 195–99, 201–8, 211, 247, 307
Sahtouris, Elisabet, xix, 56, 93–95, 101-102, 156, 173, 274
Sandage, Alan, 10, 129
sandhill cranes, 192
San Luis Valley, 191
schaumkommen, 63
Schlosshauer, Maximillian, 72
Schrödinger, Erwin, xv, 40, 63
Schrödinger's Equation, 40, 69
science
  reductionist, 38
  traditional, 14
scientific worldview, 29
scientists, materialist-based, 38
Scoles, Sarah, 75-76
sea otters 188
sedimentary layers, 168
self-creation, 157, 170, 174, 218
  cellular, 176
self-determiners, 90
self-duplicating, 222–23
self-identity, 259
self-maintenance, 174, 176

CPSIA information can be obtained
at www.ICGtesting.com
Printed in the USA
JSHW021852201219
3132JS00001B/1

9 781977 219459